Kaizen Event Fieldbook

Kaizen Event Fieldbook

Foundation, Framework, and Standard Work for Effective Events

Mark R. Hamel

with foreword by Arthur Byrne

Copyright © 2010 Society of Manufacturing Engineers

987

All rights reserved, including those of translation. This book, or parts thereof, may not be reproduced by any means, including photocopying, recording or microfilming, or by any information storage and retrieval system, without permission in writing of the copyright owners.

No liability is assumed by the publisher with respect to use of information contained herein. While every precaution has been taken in the preparation of this book, the publisher assumes no responsibility for errors or omissions. Publication of any data in this book does not constitute a recommendation or endorsement of any patent, proprietary right, or product that may be involved.

Library of Congress Catalog Card Number: 2009934639

International Standard Book Number: 0-87263-863-4, ISBN 13: 978-0-87263-863-1

Additional copies may be obtained by contacting:
Society of Manufacturing Engineers
Customer Service
One SME Drive, P.O. Box 930
Dearborn, Michigan 48121
1-800-733-4763
www.sme.org/store

SME staff who participated in producing this book:

Kris Nasiatka, Manager, Certification, Books & Video
Rosemary Csizmadia, Senior Production Editor
Frances Kania, Production Assistant

In memory of Bill Moffitt.

For the benefit of the willing student—read, but learn, first and foremost, by doing!

In sincere gratitude to the sensei who have enriched my life, my wife, Mary Ellen, my children, Jack, Kate, and Molly, and Mary Immaculata.

BOOK REVIEWER COMMENTS

"The Fieldbook is a must read for those wanting to understand the kaizen methodology. Mark Hamel clearly identifies its role utilizing the TPS version of SDCA and PDCA. He further traces his follow-through model to the 'learn-by-doing' methodology of the Training Within Industry (TWI) program, which was developed in the U.S. during WWII, and is going through a resurgence today as companies struggle to sustain kaizen gains. This book now has a prominent place on my bookshelf."

—Robert J. Wrona, Executive Director, TWI Institute®

"A critical distinction from other texts on the market, the Fieldbook links the technical aspects of kaizen to lean philosophy. Kaizen is the game changer in any lean transformation, and this is the game book."

—Bruce Hamilton, President, Greater Boston Manufacturing Partnership

"Executing effective kaizens is always a challenge. I have not seen any other book on the market that gets into this level of understanding kaizen events—from pre-planning to follow-through. This book will benefit not only the new adopters of lean, but also the early adopters who require a 'back-to-basic' understanding of how to properly execute kaizen events to effectively drive change."

—Richard Levesque, Vice President and General Manager, Professional Division, MAAX U.S. Corporation

"The Fieldbook is a roadmap for organizations to follow as they actively seek a continuous improvement culture and positive bottom-line performance. It is an essential reference for the proper application and implementation of kaizen. I intend to provide everyone on my management team with a copy."

—Max Willsie, Plant Manager, Toyota Boshoku Canada, Inc.

"Applicable to any industry, the Fieldbook shows you what kaizen looks like when done properly. I plan to give one to each of my CEO clients."

—Linford Stiles, Chairman and CEO, Stiles Associates

"Lean is about results and outcomes, not just intense focus on the process. The *Kaizen Event Fieldbook* balances theory with the tactical nuts and bolts, providing a practical roadmap for managers at all levels to strategically deploy kaizen and bring positive results to the bottom line."

—Ted Gramer, Executive Vice President and General Claims Manager, Liberty Mutual Group

"An exceptional and in-depth review of the technical components of kaizen, I found this book to offer sound guidance on how to avoid the pitfalls of inadequate preparation, uncommitted leadership, and lack of focus on sustaining the improvements in business performance that lean anticipates. It is a true reference that I anticipate returning to again and again."

—Jack M. Dutzar, M.D., President/CEO, Fallon Clinic, Inc.

"Amazingly prescriptive and reflecting years of knowledge acquired from renowned sensei, Mark Hamel has done a masterful job with the Fieldbook. Managers, instructors, and employees in any lean or aspiring lean company, within any industry, will find it an indispensable reference to impact positive, sustainable change. I plan to use it as a teaching tool for employees throughout my organization."

—David A. Amrhein, Vice President, Operations and Lean Enterprise, Ascent Healthcare Solutions, Inc.

"During any lean implementation, the pace of kaizen events is critical to maintaining momentum. This well organized, readable book will help us standardize and refine our approach."

—Kenneth Chandler, Vice President of Operations, Smith & Wesson

"Engaging your entire workforce in driving business results is fundamental to lean success. Mark Hamel has done an outstanding job of laying out the team kaizen methodology. The Fieldbook is a practical guide and I recommend it as a read for every business leader and lean practitioner."

—Jerome D. Hamilton, Global Director, Lean Six Sigma & Business Initiatives, 3M Industrial & Transportation Business

"In the *Kaizen Event Fieldbook* Mark Hamel reveals the 'tricks of the trade' for leading effective kaizen events. More than just the technical tools, he emphasizes the importance of understanding and mastering the emotional element to influence and engage employees at every level, which is critical to creating a lean culture."

—Julie DeWane, VP Global Supply Chain, GE Security

"Individuals and organizations looking to either get started or improve their kaizen capabilities will appreciate the hands-on, step-by-step approach provided in the Fieldbook. A guide for success, it also warns of the pitfalls to watch out for on the course of the journey. I am definitely ordering this book for my entire team and selected leaders, and would recommend it to anyone serious about lean."

—Stephen R. Malick, VP, WW Business Improvement, Johnson & Johnson Vision Care, Inc.

CONTENTS

About the Author xi
Acknowledgments xiii
Foreword xv

Part I—Foundation and Framework

1 Getting Started 3
   Kaizen Event Effectiveness: Prerequisite
     for Lean Transformation 3
   Audience 4
   Purpose of the Fieldbook 8
   What the Fieldbook is Not 9
   How to Apply the Fieldbook 10
   Fieldbook Structure 10
   Summary 12

2 A Short Course in "Kaizenology" 13
   Kaizen Defined 13
   Muda and His Two Brothers 13
   The Heritage of Kaizen 16
   A System Perspective 21
   Tool-, System-, and Principle-driven
     Kaizen 28
   Kaizen Event Pull 29
   Two Basic Levels of Kaizen 31
   The Lean Business Model: Context for
     All Kaizen 32
   The Kaizen Event: Rapid Improvement
     Vehicle 36
   Summary 36

3 Transformation Leadership 39
   Leadership Trumps All 39
   Technical Scope 40
   The Lean Performance System 43
   Lean Transformation Leaders 52

   Emotional Scope 54
   A Transformation Leadership Model 61
   Summary 64

Part II—Standard Work: The Multi-phase Approach

4 Strategy—Right Wall, Right Ladder 69
   Phase 1: Strategy 69
   Strategy Deployment 72
   Long-term Scheduling 78
   Summary 82

5 Plan for Success 85
   Phase 2: Pre-Event Planning 85
   The Four Planning Sub-processes 86
   Event Selection and Definition 88
   Communication 108
   Pre-work 111
   Logistics 116
   Summary 117

6 Event Execution 121
   Phase 3: Event Execution 121
   Kick-off Meeting 123
   Pre-Event Training 127
   Kaizen Storyline 132
   Team Leader Meetings 155
   Work Strategy and Team Effectiveness 158
   Report-Out 172
   Recognition and Celebration 175
   Summary 176

7 Follow-through 179
   Phase 4: Event Follow-through 179
   Event Management Improvement 188

Communication 192
Record Retention 193
Summary... 194

Part III—Developing Internal Capability: The Lean Function

8 The Kaizen-ready Enterprise 197
   Making Kaizen Your Own 197
   Kaizen Promotion Office 200
   Kaizen Promotion Officer Selection 212
   Next Steps 216
   Summary... 217

Bibliography 219
Glossary... 221
Appendix A: Blank Forms 229
Appendix B: Daily Kaizen—Beyond the Event ...249
Index... 255

ABOUT THE AUTHOR

Mark R. Hamel is a lean six-sigma implementation consultant. He has played a transformative role in lean implementations across a broad range of industries including aerospace and defense, automotive, building products, business services, chemical, durable goods, electronics, insurance, healthcare, and transportation services. A successful lean coach to leaders and associates, he has facilitated hundreds of kaizen events and conducted numerous training sessions and workshops.

Mark's 19-year pre-consulting career encompassed executive and senior positions within operations, strategic planning, business development, and finance. His lean education and experience began in the early 1990s when he conceptualized and helped launch what resulted in a Shingo-award-winning effort at the Ensign-Bickford Company.

Mark holds a BS in Mathematics from Trinity College in Hartford, Conn., a MS in Professional Accounting from the University of Hartford, and a MA in Theology from Holy Apostles College and Seminary. He is a CPA in the state of Connecticut and is dual APICS: The Association for Operations Management certified in production and inventory management (CPIM) and integrated resource management (CIRM). A national Shingo Prize examiner, Mark assisted in the development of the Society of Manufacturing Engineers/Association for Manufacturing Excellence/Shingo Lean Certification exam questions. He is also Juran certified as a six-sigma black belt and a member of the Society of Manufacturing Engineers, Association for Manufacturing Excellence, and APICS: The Association for Operations Management.

Mark can be contacted at mark@kaizenfieldbook.com.

ACKNOWLEDGMENTS

"Whoever ceases to be a student has never been a student."
—George Iles

One of the lean community's defining characteristics is learning by seeing, doing, and studying. Lean thinkers and practitioners are perpetual students. Those who teach humbly recognize that they cannot know it all and that there is so much to learn. I readily consider myself in that category.

A *sensei*, Japanese for teacher, more literally, "one who has gone before," is not "hatched." Early on, every sensei needed his own sensei to instruct, to cajole, to challenge, to allow him to fail, and to evoke self-reflection and adjustment. My sensei included the late Bill Moffitt and his colleague Bob Pentland. To those I add many others—colleagues, friends, and clients to whom I would like to express my sincerest thanks. While I will refrain from identifying them here in fear that I will overlook someone, I would like to recognize the following people who have generously contributed their time and knowledge to make this book better. I have learned, as I trust the reader will, from what they have shared.

- Matthew Ayers
- Michael Bailey
- Jill R. Behringer
- Edward P. Beran
- Evan M. Berns*
- Ginger Chandler
- Carl M. Cicerrella
- James J. Cutler
- Jerry C. Foster
- Harold V. Hassebrock
- Richard A. Jeffrey*
- Robert L. Klesczewski
- Joseph I. Murli
- Michael O'Connor, Ph.D.*
- John A. Rizzo
- Craig Robbins
- Bruce E. Thompson
- Charles J. Wolfe

*Evan, Richard, and Michael spent many hours reviewing early drafts. For that I am most grateful.

I would also like to recognize Rosemary Csizmadia, SME Senior Production Editor, who patiently coached this neophyte author and helped make this work more readable than it would have been.

FOREWORD

I began my own lean (kaizen) journey in January of 1982 when I became general manager of the High Intensity and Quartz Lamp Department of the General Electric Company. We initiated a simple kanban pull system between my department and one of my suppliers who was also a part of GE's Lighting Business Group. My inventory of quartz arc tubes (the initial kanban target) dropped from 40 days to 3 days and my supplier department totally eliminated the inventory of these parts. The side benefits were so significant that I was hooked: space freed up, better customer service, better quality, and higher productivity. Implementing the Toyota Production System in all the businesses I ran after that became my priority.

I wasn't exposed to the Toyota kaizen event approach, however, until after I left GE at the end of 1985 when I joined the Danaher Corporation where I was a group executive responsible for eight of its then 13 companies. In the summer of 1987, Danaher became the first U.S. client of the recently started (1986) Japanese consulting company, Shingijutsu. All three of the Shingijutsu founders (along with the fourth initial partner who joined them a year later) had spent their entire careers at Toyota. For a number of years prior to founding Shingijutsu, they worked directly for Taiichi Ohno, the father of the Toyota Production System, implementing TPS in the Toyota group companies and in the tier one supplier group. Their initial kaizen efforts focused on two of my companies, Jacobs Engine Brake (Jake Brake) Company and its sister division, the Jacobs Chuck Company. The results at both were amazing.

They both were so bad when we started, however, that we couldn't get Shingijutsu to work with any of our other businesses until they were comfortable that Brake and Chuck were far enough along on their lean journeys. After several years we got them to branch out into a couple of my other businesses, but that still left the bulk of the Danaher divisions without any lean help. To solve this problem, we instituted what we called the "President's Kaizen" to spread the knowledge and speed the implementation throughout the rest of the company. Every six weeks, a series of kaizen events were held in a Danaher plant where the kaizen team members were the 13 division presidents and their vice presidents of operations. We made a lot of progress and had a lot of fun along the way.

In the end, lean (the Danaher Business System) became the way of life for Danaher and resulted in a price-to-earnings multiple that was double that of other similar companies. This, plus a lean culture, has allowed Danaher to be one of the best-performing public companies over the past 20 years.

In September of 1991, I took the lessons learned and became CEO of the Wiremold Company headquartered in West Hartford, Conn. and, as they say, the rest is (lean) history.

While kaizen events are an extraordinary continuous improvement delivery and deployment mechanism, it is important to make the distinction that they do not equal lean. In fact, many people mistakenly and myopically believe that because they are doing kaizen events, they are "doing" lean. This couldn't be further from the truth. Lean transcends tools, events, and even systems. The Fieldbook recognizes this reality and, at the same time, properly asserts that, "a company that does not possess and routinely exercise the capability to effectively target, plan, execute, and follow through on their kaizen events, including the non-negotiable requirements to comply with the new standard work to sustain the gains, cannot and will not successfully transform themselves into a lean enterprise." In other words, get good at kaizen and sustaining the gains, or forget about lean.

It is for this very reason that I, as Wiremold CEO, *personally* trained hundreds of employees in lean principles and then *personally* facilitated dozens and dozens of kaizen events. On the heels of the launch, we made use of the best sensei to accelerate and expand lean learning and application through kaizen events. One of the sensei, the late Bill Moffitt, was my friend and one of Mark Hamel's teachers.

Wiremold quickly supplemented the sensei approach with the development of an internal lean function. The people, the true heart and soul behind any lean transformation, responded. The company's overall improvements approximated the typical kaizen event results. For example, from the end of 1991 until the middle of 2000:

- delivery lead time was compressed from 4–6 weeks to 1–2 days,
- product development lead time dropped from 2 years to 3–6 months,
- space requirements were halved,
- inventory turns increased from 3 times to 18 times,
- productivity improved by 162%,
- customer service satisfaction went from 50% to 98%, and
- the gross profit percentage gained 13 points.

The enterprise-wide improvements drove a 13.4 fold increase in operating profit while sales growth, aided by acquisitions that were enabled by drastic working capital reductions, resulted in Wiremold being able to double sales twice (once every four years). More importantly, we were able to take a company that was valued at $30 million at the end of 1990 and sell it for $770 million in the summer of 2000 (a gain of 2,467%).

These types of results may seem other-worldly, but they need not be. Breakthrough performance is achievable no matter the industry or value stream. Lean is not, as was initially claimed by the naysayers, limited to the automotive industry; nor is it only a manufacturing "thing." Due to its universal principles and common enemies (waste, unevenness, and overburden), lean and with it, kaizen, extends to healthcare, financial services, insurance, government, transportation, entertainment, etc.

Mark Hamel is a lean teacher, a sensei, who has studied under some of the very best and has learned by doing. He has the mind and motivation of a learner, as evidenced by his "arrangement" (through a colleague's wife who happened to work for me at Wiremold) of a 1994 presentation by yours truly at his employer. That presentation sparked a Shingo-Prize-winning lean transformation and launched his immersion into lean thinking, practicing, and teaching. I have had the benefit of Mark's lean implementation expertise at several of my business interests over the years.

The full name of his work, *The Kaizen Event Fieldbook: Foundation, Framework, and Standard Work for Effective Events*, like the page count, is pretty big. It's big, but very important. The Fieldbook should be required reading for any enterprise that truly seeks to become lean. It necessarily addresses the basic foundation of lean and the roots of kaizen. Further, it briefly explores something very near and dear to my heart, lean leadership. And, it lays out the standard work for how to "pull" kaizen by recognizing strategic imperatives and value stream improvement needs, and then rigorously planning, executing and following through on each kaizen event. Finally, the Fieldbook provides insight into how best to establish your own lean function or kaizen promotion office.

In short, for those who are profoundly committed, lean transformation efforts are absolutely worth the necessary blood, sweat, and tears. Ineffective kaizen events, however, induce disproportionate and unnecessary suffering and can quickly derail a lean implementation. The prize goes to those who understand that much of lean is about working smarter, not harder. The Fieldbook positions both the practitioner and the lean leader to work a lot smarter and a lot more effectively.

What now? Go to the gemba and put the Fieldbook in action!

Arthur Byrne
Operating Partner
J.W. Childs Associates, L.P.
Retired Chairman, President, and
  CEO of the Wiremold Company
Avon, Conn.
September 8, 2009

PART I

FOUNDATION AND FRAMEWORK

1
GETTING STARTED

KAIZEN EVENT EFFECTIVENESS: PREREQUISITE FOR LEAN TRANSFORMATION

Over the past decade, there has been an explosion of lean studies. The types of media and delivery vehicles that have been deployed are many, including books, magazine articles, white papers, conferences, seminars, certifications, "belts," and blogs. The subject matter spans from the elemental with a focus on specific tools and techniques to the more holistic where the emphasis is more systemic and enterprise-wide.

Yet, despite the proliferation of lean "knowledge" and lean activity at the *gemba* (the Japanese term for the "actual place" where the work is done), many companies are still struggling to find their way. A recent study reflects that 59% of the 2,500 business people surveyed were either in the planning or early stages of lean, as contrasted with the 7% allegedly enjoying an "advanced" level of lean implementation and 34% at "extensive." The same survey respondents identified their company's primary barriers to the creation of a lean enterprise as:

1. pushback from middle management (36%),
2. lack of implementation know-how (31%),
3. employee resistance (28%), and
4. supervisor resistance (23%) (Lean Enterprise Institute 2007).

All of these barriers, except the second, are of a change management and transformation leadership nature. However, it is safe to say that the remaining barriers are, directly and indirectly, partly addressable by the effective removal of the second barrier. Competency and success are effective countermeasures to many of the root causes of resistance.

One of the most significant components of implementation know-how is the kaizen event. While the engine of any lean transformation is people, the kaizen event is a primary vehicle for engaging the hearts and minds of the workforce. Kaizens teach people how to see, think, and feel within the context of lean and, ultimately, how to rapidly and effectively deploy their improvement ideas to address high-impact opportunities.

In fact, there is a lot of talk about material and information flow within the scope of value stream mapping specifically, and lean in general. Kaizen is much about idea flow—first getting the ideas, big and small, to flow within a scientific construct and then ultimately implementing them and sustaining results. In the end, it is

Over 35 real life stories are interspersed throughout the Fieldbook. These "Tales from the Gemba," or "Gemba Tales" for short, are based upon the experiences of the author and other lean practitioners. The stories provide insight into the application of the concepts. Sometimes they reflect success and other times failure, but the intent is always to teach and, if possible, share some gemba humor.

> **Kaizen**
>
> The word "kaizen," Japanese for continuous improvement, and more formally defined in Chapter 2, is used as both a noun and a verb. As a noun, "kaizen" is shorthand for a kaizen event or a smaller discrete kaizen activity, while kaizens (plural) is for multiple kaizen events. As a verb, it represents the action of continuous improvement.

about human return on ideas. Of course, one central and pragmatic tenet is that kaizen must be done *with* people and not *to* people. The kaizen (and lean) foundation is built upon humility and respect for the individual.

Some may judge this a bold statement, but . . . *A company that does not possess and routinely exercise the capability to effectively target, plan, execute, and follow-through on its kaizen events, including the non-negotiable requirement to comply with the new standard work to sustain the gains, cannot and will not successfully transform into a lean enterprise.* Kaizen event effectiveness is thus a prerequisite for lean transformation success and a precursor of a daily kaizen culture. Due to the dynamics of transformation leadership, the converse is not necessarily true; a company can be good at kaizen events, but still fail in its effort to become lean.

The scope of this book is primarily limited to kaizen standard work and event management. It is focused on imparting the reader with the strategies, tools, and techniques necessary to conduct effective kaizen events, with some forays into other supporting elements of the lean enterprise. However, before beginning, it is useful to understand some of the symptoms of ineffective kaizen events. Table 1-1 will help the reader discern whether there indeed may be a kaizen performance gap within his or her organization.

During the outset of a lean launch the symptoms outlined in Table 1-1 are imperceptible to all but the most experienced. The symptoms are often masked by the initial energy and euphoria that accompany any new initiative, and the first flurry of activity and the easy elimination of the most egregious forms of waste. To some degree, this is to be expected; people are at the beginning of the learning curve. However, kaizen event malpractice will eventually limit the success of the fledgling transformation.

Flawed kaizen management approaches sap the vitality out of any lean transformation effort. Effective lean leaders employ and propagate best practices, "standard work" in lean parlance, to limit malpractice and dramatically increase the odds of success.

Standard work, also known as standardized work, is a fundamental lean tool that explicitly defines and communicates the current best practice (least wasteful) for a given process that is dependent upon human action. It provides a routine for consistency, relative to safety, quality, cost, and delivery, and serves as a basis for improvement. Standard work is comprised of three basic elements:

1. *takt* time (rate of customer demand),
2. work sequence, and
3. standard work-in-process.

The notion and relevancy of kaizen event standard work is addressed throughout this book.

AUDIENCE

This book is written primarily for lean practitioners, both existing and aspiring, whose population is comprised of three groups (see Figure 1-1):

1. kaizen technologists and facilitators in the kaizen promotion office,

> **The Universality of Kaizen**
>
> Kaizen "works" in every industry and, by extension, every enterprise. This can be deduced from vast empirical evidence and corroborated by the experiences of lean practitioners far and wide. So why is kaizen so universal? Every value stream is comprised of processes which, in turn, are comprised of activities. Virtually every activity has a measure of waste, some a lot, some a little. Kaizen does not discriminate, waste is waste. Kaizen's very core is about identifying and eliminating waste with a vengeance and implementing management systems to ensure sustainability of the improvements.

Table 1-1. Nine symptoms of kaizen event malpractice

Symptom	Description	Root Causes (Related Book Chapters)				
		Leadership (3)	Strategy (4)	Planning (5)	Event Execution (6)	Follow-through (7)
1. Variation	Fundamental kaizen event content, sequence, and approach vary event to event. Kaizen events are not conducted in accordance with standard work or standard work is not defined.				X	X
2. Insignificance	Events are not strategically selected to address high-impact opportunities.	X	X	X		
3. Incoherence	Events are not synergistic with other achieved or anticipated improvements within the value stream(s).	X	X	X		
4. Blindness	The kaizen event team's ability to identify waste/opportunities during the event is limited due to inadequate preparation, training, tools, or approach.	X		X	X	
5. Indifference	The kaizen event team's readiness to acknowledge identified waste/opportunities is limited by behaviorally and culturally induced "filters."	X			X	
6. Lethargy	The kaizen event team and support functions' sense of urgency is inadequate to eliminate acknowledged waste during the event.	X		X	X	
7. "Unsustainability"	Kaizen gains are not sustainable due to lack of validation, rigor in standard work development, training, follow-through, and/or day-to-day discipline.	X			X	X
8. Demoralization	Kaizen team members and others in organization are demoralized due to lack of meaningful and sustainable results, inadequate planning, resources, empowerment, recognition, follow-through, use of kaizen to "cut heads," etc.	X	X	X	X	X
9. Presumption	Kaizen events are narrowly presumed to equal lean in total and are not viewed as only part of the total lean transformation journey.	X	X			

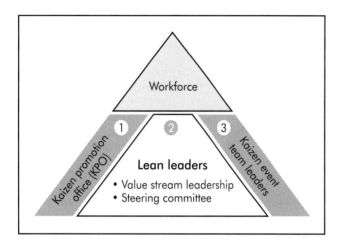

Figure 1-1. Audience of lean practitioners.

2. lean leaders, and
3. kaizen event team leaders.

Kaizen event participants may also benefit by reading the Fieldbook.

Kaizen technologists and facilitators. While the use of the term "technologist" may appear curious, it is appropriate in that it reflects someone who applies scientific knowledge to solve practical problems. (This is in purposeful contrast to "technocrat," which smacks of elitism—a very un-lean characteristic.) The science of kaizen is real and, at its core, is founded upon the profoundly simple, but powerful plan-do-check-act (PDCA) cycle. Also, it should be acknowledged that the use of technology, especially in the change management business of lean, does not preclude a necessary balance with behavioral skills.

Depending on the company, kaizen technologists and facilitators represent individuals whose titles combine words such as "vice president (VP)," "director," "officer," "manager," or "coordinator" with words or phrases like "kaizen promotion," "continuous improvement," "kaizen," "lean," "just-in-time (JIT)," "lean six sigma," or "operational excellence." There is an opportunity for a veritable title bingo card, but you get the point.

For the sake of simplicity, the title "kaizen promotion officer" will be used as the proxy for the kaizen technologist/facilitator. These individuals comprise an enterprise's "lean function," which is discussed in detail in Chapter 8. They are the organization's identified lean subject matter and deployment experts whose primary role is to teach, advise, and facilitate at all levels within the company. The placement and reporting relationships of kaizen promotion officers within the organization should be such that they can have meaningful access to others and exert the requisite amount of influence within the organization as indicated in Figure 1-2. Similarly, there should be sufficient resources, from the perspectives of the corporation, business units and value streams to drive meaningful change—through formal kaizen events, training, and projects.

Ultimately, it is the corporate kaizen promotion officer's responsibility to develop and maintain the standard work for kaizen event management. Each kaizen promotion officer, no matter the level in the organization, should be expected to adhere to the standard work and contribute to its continued improvement by way of formal activities such as periodic Lean Summit meetings and informal collaboration with fellow kaizen promotion officers and management. This subject is addressed in Chapter 8.

Lean leaders. Without the lean leaders, there is categorically no chance for a sustainable lean transformation. They provide vision, direction, urgency, resources, and accountability, and back it up with tenacious and consistent involvement and communication. Lean leadership is addressed explicitly in Chapter 3 and implicitly throughout this book.

The lean leader population encompasses all of the senior executives, business leaders, and managers, especially those who participate on a lean steering committee and who lead or manage the various value streams and key processes. Optimally, they will use lean within the framework of a lean business system and drive high-leverage kaizen events to fulfill strategic imperatives and execute the value stream improvement plans.

The various senior change agents can also benefit from a deeper understanding of kaizen and kaizen event management. This is particularly important if they are integral in the selection of kaizen promotion officers and kaizen event team leaders.

Kaizen event team leaders. A quick review of this book will provide team leaders with greater

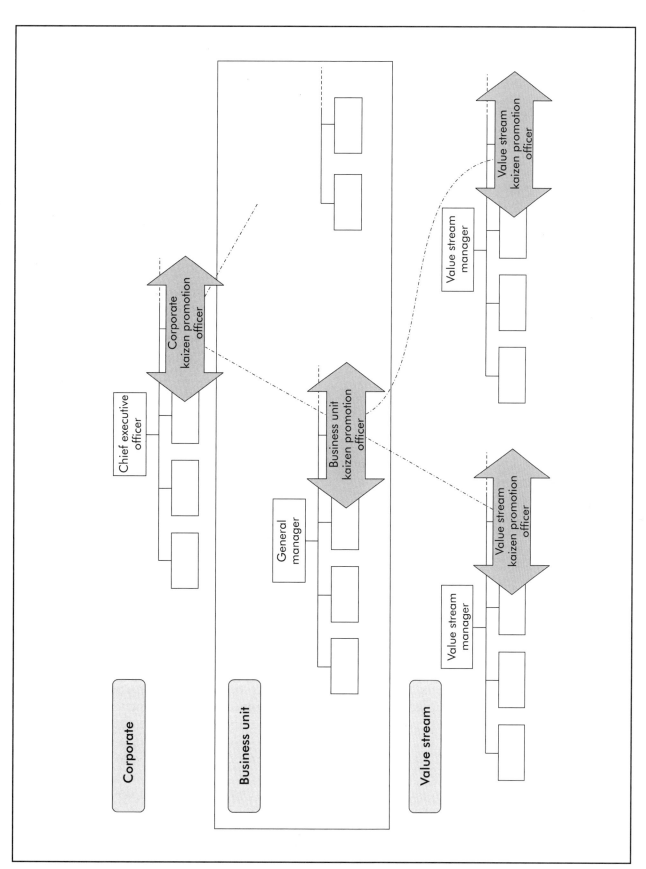

Figure 1-2. Placement and reporting relationships of kaizen promotion officers within the organization.

insight into the big picture of the kaizen event process and the underlying scientific method and strategy. Additionally, it will instill a deeper appreciation of the need for effective planning and follow-through.

PURPOSE OF THE FIELDBOOK

The Fieldbook was created to answer the needs of both the newly engaged and the experienced lean practitioner. The primary intent is twofold: 1) to provide insight into the foundational elements of kaizen and its context within a holistic lean business system, and 2) to provide the practitioner with kaizen event standard work. This is based upon the approach and philosophy employed and propagated by Shingijutsu, a self-described world consultancy for manufacturing technology founded by several former members of Taiichi Ohno's Toyota Production System implementation team. It was instrumental in the first U.S. lean transformation (Danaher's Jacobs Vehicle Systems, a.k.a. "Jake Brake," in Bloomfield, CT) and many others since.

The practitioner, unless at the absolute beginning of personal exposure to lean, undoubtedly will have had standard work, its benefits, tools, and techniques drilled into him or her by a *sensei*. "Sensei" is the Japanese term for teacher. From the perspective of kaizen event facilitation, many believe that a person only begins to *approach* proficiency after coaching 100 kaizen events. Accordingly, it is often presumptuous to self-adopt the title of "sensei." Even in Japan the term has experienced title inflation, meaning that anyone who teaches anything is apt to call himself a sensei (hence, some lean teachers have moved to the formal or informal title of "coach"). In any event, caution should be exercised when it comes to sensei selection. This role is, after all, an extremely critical element in any successful lean transformation.

Consistent with the lean "foundation" instilled by the sensei, the practitioner has hopefully enjoyed some measure of success implementing standard work on the plant floor, in the office, field, and lab, and anywhere else waste hides. So, if standard work is so powerful and important . . . and it is, why is there a dearth of standard work in the public and private domain on how to conduct an effective kaizen event?

There is standard work that most sensei do not explicitly share. The apparent lack of sharing is usually not because the sensei is being contrary. Indeed, his job, vocation even, is to teach the willing student. It is just that virtually all of the tools, techniques, and strategies are intuitive to him. In other words, the sensei *just does it*.

If the average sensei was asked how he or she facilitates an event, the reply likely would be that it is something that is learned, first and foremost, by *doing*. Indeed, the meaning of sensei is "one who has lived before." There are no shortcuts. Simply telling "how" is dangerous in that it artificially shortens the student's developmental journey. This can rob the student of the natural growth inherent in the pure experience of doing and reflecting, which is best supplemented by a sensei's Socratic Method. Of course, not every corporate environment has the requisite patience or readily available sensei. Nevertheless, if the sensei were to entertain such a question, he or she would likely have to stop and think. The sensei would recount the process, decisions, tools, and strategies, and then provide a perhaps esoteric response. The "answer" is most difficult because, while many elements of the strategy, pre-event planning, and follow-through phases are fairly straightforward, there are numerous details that must be considered.

The actual event execution has many straightforward elements. However, a lean sensei will tell you that no two events are ever exactly the same. There will always be one or many more differences in the scope, targets, process, tools and, without question, team dynamics. Accordingly, the sensei often *feels* his way through the event, introducing new tools or techniques at the appropriate time, pushing teams one way or another, altering strategy, etc. So, it is hard, if not impossible, to capture this dynamic within standard work. This leads to another reason why there is little in the way of substantive kaizen event standard work—it is downright painful to articulate and, in some ways, dangerous to attempt such an explicit summary.

A serious student of kaizen event management must never forget that learning comes from direct observation of the sensei, instruction from the sensei, studying and, most importantly, facilitating and participating in kaizen events. As in anything, learning comes from both success and failure. As one person was wont to say, "Knowledge makes a bloody entrance." Perhaps less graphic is the sentiment that "Wisdom grows out of difficulties" (Hirano 1990).

WHAT THE FIELDBOOK IS NOT

As reflected in the discussion thus far, the Fieldbook is many things. However, to appropriately set expectations, it is important to articulate what *it is not*:

- *The "Lean Gospel."* The Fieldbook is not intended as a compelling exhortation to undertake a lean transformation. It presumes that the reader has taken the leap of intellect and faith to the lean side. The Fieldbook does not attempt to point out the errors of batch-and-queue thinking, the wiles of ill-conceived automation, or the corrosive effects of poor workplace organization. It does not try to contrast the key performance indicators or measures of market value added in lean and non-lean companies. Nor does it try to convince the reader to vigorously implement continuous flow wherever possible or apply autonomous maintenance or balance work content among multiple operators.
- *A "how to book."* The Fieldbook is not a treatise on the how, where, and when to implement any of the various lean tools and techniques—5S (sort, straighten, shine, standardize, sustain), visual controls, continuous flow, kanban, heijunka, single-minute exchange of dies (SMED), mistake proofing, total productive maintenance (TPM), jidoka, strategy deployment, value stream mapping, lean management systems, etc. However, it does presume that the reader already has, at a minimum, an understanding of lean tools and techniques, their interrelationships, leverage, primacy, and sequence of use.

Further, the Fieldbook is not a cookbook. It must be understood that the notion of standard work is treated in this book at the conceptual level. To successfully facilitate a kaizen, a facilitator must possess a mix of technical and behavioral skills that enable him to transcend a paint-by-numbers approach (otherwise, everyone could do this!). So, it may be helpful to think of the Fieldbook as a set of standard work *and* guidelines.

- *A substantive exploration of daily kaizen.* Daily kaizen is the genuine implementation of continuous small incremental improvements, which is representative of a profoundly engaged workforce that is empowered to experiment and make things better every day. The Fieldbook is unabashedly focused on kaizen events that are more closely aligned with breakthrough performance and are typically the initial laboratory in which employees learn about kaizen. Events dynamically support and coexist with daily kaizen as the lean transformation matures. (Appendix B devotes several pages to the subject of daily kaizen.)

There are a multitude of excellent books that address the broader topic of lean, the required cultural shifts, and the various specific tools and techniques so necessary to a successful lean transformation. It would be useful to obtain and peruse the recommended reading list for Lean Certification (available from the Society of Manufacturing Engineers, www.sme.org/certification). Also, the Glossary herein contains brief definitions of many of the lean tools.

The best practices embodied in the Fieldbook can be characterized within the three basic elements of standard work:

1. *takt time*, which is a measure of the rate of customer demand (takt is a German word that means meter and is calculated by dividing the available time of a given process for a given period by the customer demand for that period);
2. work sequence, and
3. standard work-in-process (WIP).

Each event has its own implicit takt time, relative to the duration of the kaizen event and the prescribed timing of the pre-event planning and post-event follow-through. Similarly, there is a best practice sequence for planning, executing, and following up after a kaizen. Within each one of these macro processes, there is standard work for the micro-level processes. Standard WIP can be understood within the context of the progressions of tasks and the collection and analysis of data, which is guided by the many standard forms and work sheets and captured and reflected in charts, maps, etc.

Consistent with the notion of guidelines, within this amalgam of standard work, the practitioner must learn how to *think* to facilitate an effective kaizen event; effectiveness is measured by the achievement of kaizen targets and the behavioral and technical development of the participants. There is an underlying philosophy, process, and story behind kaizen (more on this in the next chapter) to which the practitioner must remain true. This does not mean that the event standard work is not applicable. Quite the contrary, it means that the facilitator must understand how to weave the standard work within the kaizen. This sort of rigid flexibility is necessary because, as mentioned before, every event is unique. There will always be some measure of variation because of dynamics introduced by changes or differences in scope, targets, takt times, process issues and technologies, improvement ideas, environments, and people. Accordingly, the value of experience and good old-fashioned common sense, combined with the time-proven lean tools and techniques, can never be underestimated.

HOW TO APPLY THE FIELDBOOK

The Fieldbook can be applied in several ways: 1) at a corporate level as a framework for developing (improving, if the "developing" is done) and deploying standard work throughout the organization, and 2) for the personal development of motivated kaizen practitioners. Many of the figures contained herein are available for download at www.kaizenfieldbook.com.

Certainly, any company heading down the lean transformation path must develop an excellent in-house capacity and competency to conduct numerous, frequent, and effective kaizens in a variety of different venues. In so many ways, the kaizen event is the initial (and repeat) lean delivery, cultural indoctrination, and training system. It is too important to risk dropping the ball in this area. Kaizen event standard work is one of the great enablers.

In the situation where an organization does not have a resident lean sensei, it is most advisable to get one quickly. Second only to finding a change agent at the most senior executive level is the need to secure the requisite knowledge (Womack and Jones 1996). The most expedient way of doing that, presuming that there is no candidate in house, is to hire a full-time sensei or consultant who will in turn develop your own in-house capabilities.

Even if there is a resident sensei, there still may be a critical need for a consultant, who should bring blunt objectivity and a wealth of experience and perspectives from a variety of applications, industries, and clients. The decision to use a consultant should be based upon several factors, including the magnitude of the performance gaps, the rapidity with which the gaps must be closed, the number and diversity of company locations and operations, and the extent of the required cultural change. It is also based upon leadership's desire for the benefits of the outsider's:

- fresh eyes,
- propensity to tell it like it is, and
- frequent challenge to achieve what may be (currently) unimaginable.

FIELDBOOK STRUCTURE

The Fieldbook's eight chapters are apportioned into three parts:

1. Foundation and Framework,
2. Standard Work: The Multi-phase Approach, and
3. Developing Internal Capability: The "Lean Function."

Part I, encompassing Chapters 1 though 3, is structured to provide the reader with an overview of kaizen from a definitional and conceptual perspective. The role and importance of transformation leadership is reviewed. Part II, Chapters 4 through 7, details the kaizen event standard work, spanning the logical sequence of:

1. strategy,
2. pre-event planning,
3. event execution, and
4. event follow-through.

Finally, Chapter 8 comprises Part III, moving from the specifics of kaizen standard work to the development and customization of the enterprise's lean function. Lastly, the Fieldbook contains a glossary and two appendices.

Part I—Foundation and Framework

Chapter 1, Getting Started. This is the chapter you are reading now. It establishes the premise that effective kaizen events are a prerequisite for true lean transformation, while specifying the Fieldbook's audience, scope, and purpose.

Chapter 2, A Short Course in "Kaizenology." In this chapter, the reader is provided with the source, foundation, and fundamentals of kaizen, its Training Within Industry (TWI) job methods, plan-do-check-act (PDCA) and standardize-do-check-act (SDCA) roots, and its context within a lean business system. The chapter is "dense" from the perspective of background, concepts, principles, systems, and tools. There may be a desire to skip to Chapter 3 or 4 and jump right into the mechanics. This may be fine if you have a deep understanding of the substance within Chapter 2, otherwise, "grind it out" and study. The risk of jumping downstream of Chapter 2 is that you will learn "how," but may miss much of the "why."

Chapter 3, Transformation Leadership. This chapter addresses what is often the elephant in the lean room that no one wants to acknowledge—poor transformation leadership. Without good leadership, kaizen is just a bunch of *muda* (waste). It addresses the basic things that lean leaders must do well:

- know, execute, and require others to adhere to the kaizen event standard work,
- apply the rigor of the lean performance system,
- develop an effective KPO function,
- effectively manage change, and
- develop their own personal lean competency.

Part II—Standard Work: The Multi-phase Approach

Chapter 4, Strategy: Right Wall, Right Ladder. This chapter addresses the first of the four-phase kaizen event methodology. The strategy, as articulated within the enterprise's strategy deployment process as well as the supporting value stream improvement plans, must be the primary driving force behind each kaizen event. Absent of this linkage, there is a real risk of conducting indiscriminate kaizen events, which is waste, by any definition.

Chapter 5, Plan for Success. This chapter sheds light on the unfortunately much overlooked and marginalized pre-event planning process:

1. event selection and definition,
2. communication,
3. pre-work, and
4. logistics.

The underlying standard work is supplemented by several important templates and forms.

Chapter 6, Event Execution. Many would consider this chapter to be the "meat and potatoes" of kaizen event standard work . . . and they would be partly correct (do not forget the other three phases!). There is a tremendous amount of important material in this chapter, including details on the seven-part kaizen event sequence:

1. kick-off,
2. pre-event training,
3. the kaizen "storyline,"
4. team leader meeting process,
5. kaizen work strategy,
6. report-out, and
7. recognition and celebration.

Chapter 7, Event Follow-through. The post-kaizen event presentation applause does not represent the official end of the kaizen, but rather the transition from one phase to another. This chapter is ultimately about sustaining the kaizen gains, improving the organization's kaizen event management capabilities, and post-event communication. It addresses elements such as the kaizen newspaper, leader standard work, the post-event audit, and the kaizen event evaluation.

Part III–Developing Internal Capability: The "Lean Function"

Chapter 8, Becoming a Kaizen-ready Enterprise. After gaining an understanding of how to conduct kaizen events, the next question is often (or should be), "How do we make kaizen our own?" For a company to become proficient at kaizen, it must move beyond "sensei dependency." The KPO development roadmap contained within this chapter provides a simple approach to developing the "lean function" and becoming a more kaizen-ready enterprise.

Glossary. The Fieldbook may contain terms and concepts that are unfamiliar to the reader. The Glossary, while not comprehensive, defines nearly 60 important terms. It may prove useful for the reader to peruse the glossary prior to beginning Chapter 2.

Appendices. Two appendices are included in the Fieldbook. The first, *A*, contains blank forms to promote the use of standard work in your kaizen events; *B* briefly discusses daily kaizen and its relationship to kaizen events.

SUMMARY

- The majority of companies seeking to become lean have yet to achieve an advanced stage of lean implementation.
- The biggest obstacles to lean are change management issues and gaps in implementation know-how.
- Kaizen must be done *with* people and not *to* people.
- Companies that do not possess and routinely exercise the capability to effectively target, plan, execute, and follow-through on their kaizen events, including the non-negotiable requirement to comply with the new standard work to sustain the gains, cannot and will not successfully transform themselves into lean enterprises.
- The Fieldbook audience includes:
  1. kaizen technologists and facilitators,
  2. lean leaders, and
  3. kaizen event leaders.
- The Fieldbook's purpose is to provide: 1) insight into the foundational elements of kaizen and its context within a lean business system, and 2) provide the practitioner with kaizen event standard work.
- The Fieldbook is not: 1) the "Lean Gospel," 2) a "how to" book or "cookbook," 3) a deep exploration of daily kaizen.
- The Fieldbook can be applied at a: 1) corporate level and 2) an individual level for personal development.
- The Fieldbook is structured to provide the reader with insight into:
  1. kaizen's source and foundation,
  2. the role and importance of transformation leadership,
  3. the multi-phase kaizen event process and,
  4. how to make the enterprise more kaizen-ready.

REFERENCES

Hirano, Hiroyuki. 1990. *5 Pillars of the Visual Workplace: The Sourcebook for 5S Implementation*. Portland, OR: Productivity Press, p. 27.

Lean Enterprise Institute. 2007. "Survey: Lean's No. 1 Obstacle can be Found in the Middle." Dearborn, MI: Society of Manufacturing Engineers, *Lean Directions*, September, available from http://www.sme.org/cgi-bin/get-newsletter.pl?LEAN&20070910&4&; Internet accessed 9/19/07. Note that the results add up to greater than 100%. Survey participants were encouraged to select all relevant barriers.

Womack, James P. and Jones, Daniel T. 1996. *Lean Thinking: Banish Waste and Create Wealth in Your Corporation*. New York: Simon & Schuster, pp. 248-255. The suggested action plan for initiating a lean transformation within a company encompasses 7 steps: "find a change agent . . . get the knowledge [find a sensei] . . . find a lever by seizing a crisis, or creating one . . . forget grand strategy for the moment . . . map your value streams . . . begin as soon as possible with an important and visible activity . . . demand immediate results . . . as soon as you've got momentum, expand your scope."

2

A SHORT COURSE IN "KAIZENOLOGY"

"Some people are so busy learning the tricks of the trade that they never learn the trade."
—Vernon Law

Caution!
This chapter contains theory. Patience will be rewarded.

KAIZEN DEFINED

The probability of finding the word "kaizenology" in any dictionary, unlike the definition for "kaizen," is exactly zero. While "kaizenology" is a fabricated word, the need to understand kaizen at a more than superficial level is real. Before rushing headlong into the mechanics of kaizen pre-event planning, event execution, and follow-through, it makes sense for the would-be kaizen event practitioner to first internalize kaizen's roots, meaning, and underlying system—primarily from an event-based perspective.

In simple terms, the Japanese word kaizen, pronounced ky-zen, means continuous improvement. Kai means "change" and zen means "for the better" or "to take apart" and "to make good," respectively. To remain true to the definition, the scale and frequency of improvement is small and repetitive, often done over a long period of time—continuously. This is consistent with the notion of daily kaizen. Within the context of the kaizen event, the scale is typically much larger and the frequency periodic—event driven. Kaizen events, especially in comparison to the concept of kaizen as many small incremental improvements, approach the definitional understanding of *kaikaku*—the Japanese term for radical improvement. To be clear, the event does not, and can not replace the truly continuous; rather, it is a powerful means to accelerate improvement, drive breakthrough performance, and serve as a training ground and launching pad for daily kaizen with which it will coexist in proper proportion as the enterprise advances down the lean implementation path. (The subject of daily kaizen is discussed in Appendix B.)

It requires a deeper look to appreciate and begin to understand kaizen, and thus the kaizen event. Kaizen constitutes multiple layers and levels—philosophy, methodology and methods, tools, and scope. But first, it is useful to review the chief protagonist (waste) and then kaizen's heritage.

MUDA AND HIS TWO BROTHERS

Muda, or waste and its reduction, is the principle theme within kaizen. Why attack waste? Because, by its very nature it crowds out and ruthlessly strangles value. Value is defined by the customer and translates into basic economics; customers only want to pay for value. Within any marketplace that has a modicum of competition and a supply that approximates demand, the customer will buy the highest quality, lowest cost, shortest lead-time goods and services. In such a situation, the customer pays for value, while the company and its other stakeholders "pay"

> **Other Definitions of Kaizen**
> 1. "Japanese for continuous improvement, a business philosophy about working practices and efficiency; improvement and productivity, and performance." (*Webster's New College Dictionary* 2008)
> 2. "Japanese term for a gradual approach to ever higher standards of quality enhancement and waste reduction, through small but continual improvements involving everyone from the chief executive to the lowest level workers." (Businessdictionary.com 2008)
> 3. "Continuous improvement of an entire value stream or an individual process to create more value with less waste." (Lean Enterprise Institute 2003)
> 4. "Philosophy of ongoing improvement: a Japanese business philosophy advocating the need for continuous improvement in somebody's personal and professional life." (*Encarta World English Dictionary* 2007)
> 5. "The Japanese term for improvement; continuing improvement involving everyone—managers and workers. In manufacturing, kaizen relates to finding and eliminating waste in machinery, labor, or production methods. See continuous process improvement." (Cox and Blackstone 1998)
>   a. "Continuous process improvement (CPI)—A never-ending effort to expose and eliminate root causes of problems; small-step improvements as opposed to big-step improvement." (Cox and Blackstone 1998)

for the waste. This payment represents the funding and investment in resources beyond the minimum required to satisfy the customer's requirements—investments in people, equipment, materials, and space.

What is not value is waste. The words "nasty" and "pernicious" describe waste, but the ugly sounding Japanese word for it, *muda,* is just about perfect . . . guttural, with little inflection, kind of like a bad grunt. For example, think about the waste of going on a "safari" to find some information that is necessary for you to complete your job. Feel like grunting now?

While there is clearly a differentiation between value-added and non-value-added work, there is a further distinction between two types of muda; cleverly called Type 1 and Type 2. Type 1 muda consists of activities that cannot be reasonably eliminated in the near-term, while Type 2 represents waste that is a prime candidate for quick elimination through kaizen.

Type 1 muda is often called incidental or auxiliary work in that it immediately supports the real value creating work. For example, an operator's job includes inserting a part into a machine. Later on the application of jidoka can effect automatic loading, but for now this incidental work must happen or the part does not get made. Type 1 muda can also extend to activities that are required because of regulatory or internal control needs. For example, it may be more "efficient" to combine the tasks of setting up vendors in the system, processing vouchers, and cutting checks within one job, but that is just poor risk management.

Shigeo Shingo is credited with articulating the seven basic categories of waste, outlined in Table 2-1. To these wastes, many have added an eighth,

> **Gemba Tales**
>
> **It is muda whether you see it or not.** A sensei once confided on how his Japanese sensei instructed, perhaps the best word is "chastised," him when he was a novice so he would understand the importance of direct observation to identify waste. After complying with an order to stand and watch the chop-saw operation for an extended period of time, he did not have the right answer to the sensei's question as to what he observed. After absorbing the insults, he moved on . . . until the kaizen report-out, several days later.
>
> In the midst of the report-out, the Japanese sensei called the novice out, instructing him to stand up and tell everyone what he observed. His answers were no better than they were before. The Japanese sensei then proceeded to recount the subject process. The chop-saw's stroke was 6 in. (15.2 cm) and the wood that the operator was cutting was 3 in. (7.6 cm), therefore it was the waste of motion of 3 in. (7.6 cm) per stroke—"cutting air." The sensei then did the "math" out loud for all to hear, "40 strokes per hour, 7 hours per shift, 3 shifts per day, 5 days per week, 50 weeks per year . . ." The novice suffered the humiliation . . . and never forgot the lesson.

Table 2-1. Seven traditional wastes

Waste Category	Production Examples	Service or Transactional Examples
1. Overproduction—production that exceeds or is made ahead of customer requirements	Upstream operation producing parts faster than the downstream operation's cycle time	Issuing hard copy or electronic reports to recipients who do not need them
2. Waiting—delays or idle time	Waiting to resume production on a machine that has incurred unplanned downtime	Waiting for another's review or approval of a transaction or file
3. Transportation—unnecessary physical or virtual movement of materials and information	Moving materials by hand, cart, forklift, etc., multiple times throughout the process	Physically moving files from department to department when only a portion of the data is needed
4. Processing—unnecessary, excessive, or incorrect processing	Removing protective skin from a product when the customer prefers it to remain on during shipment	Conducting detailed analysis on a number of factors when only a few are critical
5. Stock—materials and supplies in excess of needs	Maintaining raw material inventory that well exceeds supplier lead time	Maintaining multiple files in the work area when only one is in process
6. Motion—excessive, unnecessary, or non-ergonomic motion	Searching for tools, materials, or information	Repeatedly backtracking through previously accessed screens to enter data during a customer contact
7. Defects—rework and scrap	Fixture induces scratches that require subsequent touch-up of the product	Incomplete data collection requiring a follow-up call to the customer

often called the "waste of a person," "waste of an opportunity," "not utilizing peoples' talents," or something of that nature. The notion behind this is not personal in any unfavorable way. During a kaizen event, the team will employ direct observation, typically of the worker. This is not because the worker is wasteful, but because waste is manifested in his activities as he deals with poorly designed processes, ill-running machines, missing or defective information or material, etc., basically dealing with one or more of the seven wastes. The premise is that if the worker is forced to "work" with such waste, then his time and talents are not focused on value, but instead are being wasted.

To these seven or eight wastes, others (probably consultants) have tacked on additional ones. They include "wastes" such as saying "no," administration, technology, creativity, and space. While discussion around these is edifying, the original seven wastes should serve the lean practitioner well. Taiichi Ohno himself allegedly defined the eighth waste as, "Forgetting what we have already learned." The implication is that standard work is how waste that was previously eliminated is prevented from coming back (more on that later); or, perhaps he was being dismissive of the innovation of the eighth waste and redirecting others back to the basics.

Muda's Brothers, Mura and Muri

Some people say that trouble comes in threes. They may have stumbled upon some sort of lean wisdom. Where there is muda, it is often accompanied, even precipitated, by a lack of process stability. The purveyors of instability include *mura* and *muri*, Japanese terms for unevenness and overburden (or strain), respectively. Mura is typified by gyrating schedules that cause pulsing—starting and stopping, hurrying and idling. The fluctuating pace, as can be imagined, drives

> **Gemba Tales**
>
> **Management-induced muri.** The new cells were designed to meet the rate of customer demand (takt time) and the standard work was verified relative to work sequence, cycle times, and standard work in process, among other things. Yet, management was enamored with speed, running a winding machine at a speed much greater than takt time—an unabashed exercise in the waste of overproduction. To add insult to injury, the speeds at which management wanted to run were much greater than the winder could consistently sustain. It continually broke down. Ultimately, takt time was not satisfied until management learned to run slower to run "faster."

inefficiency and can both cause and be caused by muda.

While mura may be unavoidable, for example auto insurance claims for physical damage with ebbs and flows based upon things like the time of the year and weather conditions (in this case, the opportunity is in resource flexibility and level-loading among far-flung operations), it is often something that is self-inflicted. In another example, sales incentives often drive drastic demand peaks and valleys, as customers rush in to exploit favorable pricing and conditions or withhold orders in anticipation of the next sales program. While this strategy may lead to the fulfillment and entertainment of the sales force, it does violence to the overall value stream performance.

Muri is even more mean-spirited in its unreasonableness. It represents demand, customer driven or otherwise, which drives people, processes, and equipment beyond their capacities relative to throughput and duration. The strain often drives availability and quality issues, which in turn drive more muda, mura, and muri.

THE HERITAGE OF KAIZEN

The lineage of kaizen can be substantially traced back to two sources, both of which originated in the United States: 1) the Training Within Industry (TWI) initiative, and 2) the plan-do-check-act (PDCA) cycle (see Figure 2-1). Interestingly, these two "seeds" of kaizen were developed at almost the same time within the U.S. and then sown in Japan after World War II. TWI was established in 1940 for the purpose of increasing U.S. wartime production output, while the first generation of the PDCA cycle was developed as early as 1939 by Walter Shewhart.

Training Within Industry

TWI's effectiveness in increasing U.S. wartime production was remarkable, especially given the breadth of deployment—over 16,000 plants. By September of 1945, the percentage of plants reporting at least a 25% improvement in key performance metrics was as follows (Huntzinger 2008):

- production increased—86%,
- training time reduced—100%,
- manpower saved—88%,
- scrap loss reduced—55%, and
- grievances reduced—100%.

The Training Within Industry philosophy espoused that every supervisor had five needs (Huntzinger 2008):

1. knowledge of the work,
2. knowledge of responsibility,
3. skill in instructing,
4. skill in improving methods, and
5. skill in leading.

The first two needs were deemed the responsibility of the company, while TWI sought to help develop, by way of its services, the last three. To this end, three "J" programs were used, each employing their own straightforward, four-step process:

1. *Job instruction* (JI) to guide supervisors in effectively training workers;
2. *Job methods* (JM) to "help the supervisors to produce greater quantities of quality products in less time, by making the best use of the manpower, machines, and materials now available" (Huntzinger 2008); and
3. *Job relations* (JR) to help supervisors in the area of human relations.

The similarity between job methods and the execution phase of the kaizen methodology is

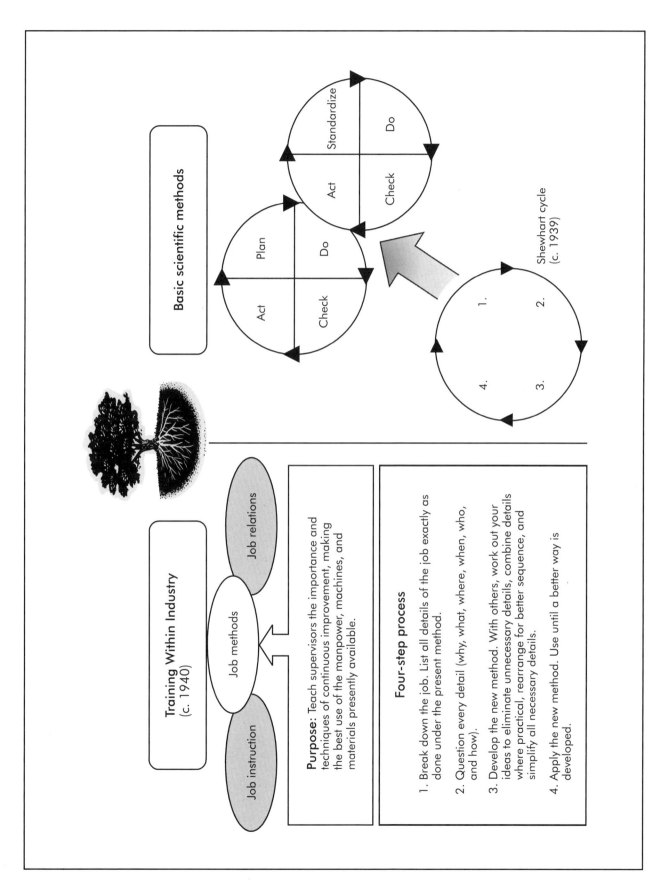

Figure 2-1. The heritage of kaizen.

striking. While the comprehensive review and analysis of kaizen execution is not conducted until Chapter 6, Table 2-2 provides some insight into the parallels. These parallels are not by chance. TWI was introduced to Japan during the Allied post-war occupation and reconstruction efforts and was absorbed throughout much of Japanese industry and especially within Toyota. Job methods, together with the PDCA cycle, are the foundation of kaizen and Japanese management methods.

The Plan-Do-Check-Act Cycle

The PDCA cycle, articulated by Walter Shewhart (see Figure 2-2), was taught in Japan by W. Edwards Deming starting in 1950. Despite Deming's reference to the concept as the Shewhart cycle, it became known as the Deming cycle or Deming wheel. This improvement cycle captures the scientific method of kaizen.

Since the time of the cycle's introduction, the step numbers have evolved such that 1 = plan, 2 = do, 3 = check, 4 = act, with 5 and 6 implicit in the continuous rotation of the wheel. In the never-ending desire to abbreviate, the cycle is now the PDCA cycle or just PDCA. Similarly, the definitions of each step can be truncated to:

1. *Plan*—establish the goals for the targeted process and identify the required changes (improvements) to achieve the goals.
2. *Do*—implement the changes.
3. *Check*—assess whether and how the executed plan delivered the intended results.
4. *Act*—depending upon the results from the prior step, standardize and stabilize the

Table 2-2. Similarities between TWI job method program and kaizen

Job Method Step	Job Methods vs. Kaizen
1. Break down the job.	The job method calls for listing all the details of the present method. Kaizen requires direct observation of the pre-kaizen situation and documentation of that reality using specific tools and forms.
2. Question every detail.	The job method prescribes asking the 5W's and 1H (why, what, where, when, who, and how) to help understand the current method and opportunities for improvement. Kaizen, at its basic level, seeks the same through the identification of root causes and the discernment of value-added activities versus waste.
3. Develop the new method.	The job method leads the supervisor to capitalize on improvements by eliminating unnecessary details, combining details when practical, rearranging for better sequence, and simplifying all necessary details. Kaizen employs these same simple strategies in the form of lean countermeasures to eliminate waste, facilitate continuous flow, etc. The job method indicates that the new method ideas should be worked out with others. Kaizen typically uses cross-functional, multi-level teams, which include the process owners. The job method calls for writing up the proposed method. Kaizen requires validating and then documenting the new standard work using specific forms/formats.
4. Apply the new method.	The job method prescribes "selling" the proposed method to the boss and the operators, getting final approval from all concerned on safety, quality, quantity, and cost, and then putting the new method to work. Kaizen, because of its collaborative team-based nature, does not need to sell. It "trystorms" and validates the new method to ensure, among other things, safety, quality, quantity, and cost. The job method requires the use of the new method until a better way is developed. Kaizen calls for following the new standard work until it is improved upon.

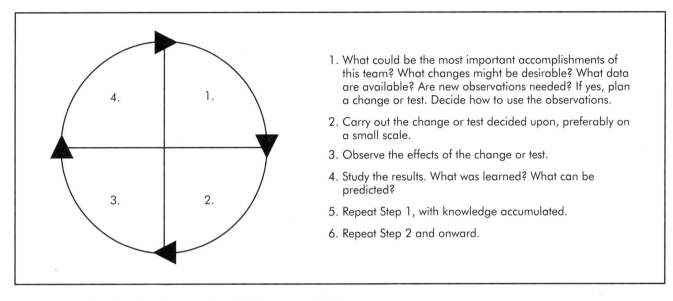

Figure 2-2. The Shewhart (Deming) cycle (Deming 1986).

improvements to sustain them or begin the cycle again.

Recently, some have renamed the fourth step as "adjust" to better reflect the step's spirit of continuous improvement. Toyota's version of PDCA uses slightly different terminology: plan-try-reflect-standardize. Its notion of standardize, in which the effective/correcting actions are stabilized, is not dissimilar to the control phase of the six sigma, define-measure-analyze-improve-control (DMAIC) methodology. Others, in perhaps an overzealous attempt to further abbreviate this scientific method, have gone to measure-improve-measure.

The Standardize-Do-Check-Act Cycle

The core scientific method of kaizen is not necessarily singular. It is not *just* PDCA. Taiichi Ohno, the father of the Toyota Production System said that there is no improvement (kaizen) without standard work. This insight, really borrowed from Henry Ford, reflects the inherent instability and un-sustainability of a condition without a standard or baseline. While the "A" in PDCA and the "S" in Toyota's PTRS do address standardization and stabilization, it is worthwhile to make this element much more explicit. SDCA or standardize-do-check-act addresses this need.

Masaaki Imai makes a careful distinction between improvement and maintenance activities; what he discerns as the two major functions of management. The two functions, including their relationship and ownership within the levels of an organization, are captured within Imai's famous kaizen diagrams (see Figure 2-3).

The left side of Figure 2-3 reflects a traditional view of improvement. To oversimplify, the (traditional) Japanese perception of "maintenance" is execution and assurance that the day-to-day operations are adhering to standard operating procedure, while "improvement" represents the elevation of the current standards. This notion of improvement is consistent with kaizen—small, frequent improvements. With equal oversimplification, the (traditional) Western perception of "maintenance" is more or less the same as the Japanese, although perhaps without the same level of discipline and rigor. However, traditional Western "improvement" leans less toward *elevation* and more toward *innovation*. Innovation is typified by substantial investment in time and money in the pursuit of the "home-run" product or process . . . and we

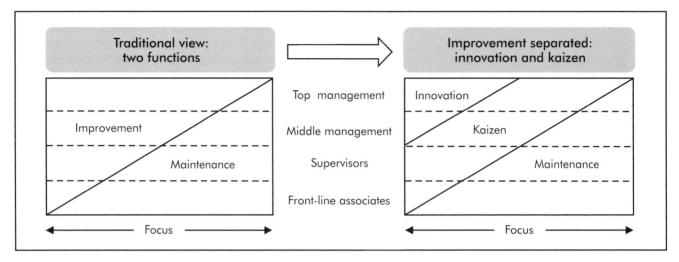

Figure 2-3. Kaizen schematic (Imai 1997).

are well aware of the batting averages of most home-run hitters.

As reflected in the right side of Figure 2-3, maintenance retains its position from the perspective of magnitude (or focus) and responsibility. However, improvement has been apportioned into kaizen and innovation. The diagonal kaizen swath indicates that kaizen is everyone's job and a quite significant one at that. Innovation is decidedly smaller (as compared to the traditional Western understanding) and is the bailiwick of only top and middle management. So, other than the important recognition relative to the focus and responsibility of innovation and kaizen, what does this mean? While innovation is outside the scope of this book, this brings us back to Ohno's assertion that there can be no kaizen without standard work.

The strict definition of *standard work*, also known as standardized work, is the established process-specific work for an operator based upon three elements:

1. *Takt time*—the rate of customer demand. The process cycle time must satisfy this rate.
2. *Work sequence*—the precise order of work steps (manual, auto, walk, and wait) that the operator must perform within the takt time.
3. *Standard work-in-process*—the requisite amount (unit count, weight, etc.) of in-process inventory at specific points within the process to facilitate flow and adherence to the work sequence.

Here we extend this definition to encompass everything in "maintenance," including that not necessarily defined by tools like standard work sheets and standard work combination sheets. This expanded scope includes things governed by visual controls (which permeate a lean environment—see Chapter 3 for a brief discussion on lean management systems), project plan performance, administrative procedures, etc. The scientific method that ensures both standardization and stability is SDCA.

According to Ohno, SDCA actually precedes PDCA (see Figure 2-4). This is because new processes, or those that have "survived" without standardization, first need to be stabilized. Lean is largely about managing the abnormal. With standardization and the assistance of simple, real-time visual controls, the abnormal is easily identified, prompting three questions:

1. Did the abnormality occur because no standard exists?
2. Did the abnormality occur because there was a lack of adherence to standard work?
3. Did the abnormality occur because the standard is not adequate?

With this insight, SDCA can be defined as follows:

1. *Standardize*—develop standards for a specific process.
2. *Do*—apply the standards.

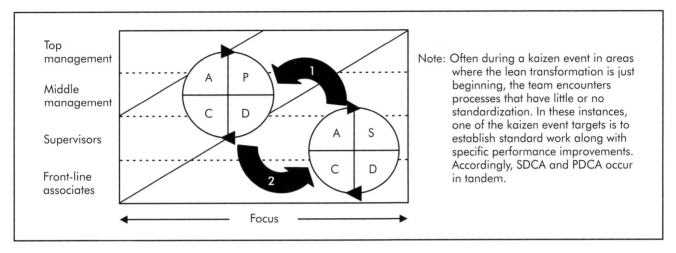

Figure 2-4. Application sequence of the scientific methods.

3. *Check*—assess whether the standards are sufficient and/or if there is a lack of adherence to the standard.
4. *Act*—depending upon the results from the prior step, make adjustments/improvements to the standardized process and/or put countermeasures in place to address the deficit in process adherence.

A SYSTEM PERSPECTIVE

The two core scientific methods of PDCA and SDCA represent the heart of what may be considered a "kaizen system." The system (see Figure 2-5) is holistic and includes an underlying:

- *Philosophy*—the system of 11 principles guide the practical and aggressive application of continuous improvement for the purpose of deploying lean within a process, value stream, or enterprise. These principles are reflected in Figure 2-6.
- *Methodology*—this is the system of the various methods applied; their order in thought or action. At least three categories have been identified:
  1. scientific (how to think),
  2. focus and alignment (where and when to apply), and
  3. deployment vehicles (how to do it).
- *Tools*—the various lean and, as appropriate, six sigma techniques employed in the methods.
- *Cultural enablers*—the organization's distinctive behavior patterns, founded upon humility and respect for the individual, facilitate and encourage continuous improvement.

Philosophy

It is a bit melodramatic to state that the unexamined lean experience is not worth doing. However, it is a good idea to explore and understand the guiding principles of kaizen.

Without over-intellectualizing it, philosophy is a set or bundle of ways of thinking, which leads to kaizen *Principle 1—Think PDCA and SDCA, the basic scientific methods*. Thinking is done with the intellect. The intellect's job, at least in theory, is to attain the truth. Truth, in the lean world, unlike the "real" world, is not relativistic or subjective; it must be based upon objective reality. In lean language, truth is called "reality," "the current state," "the current condition," or "the pre-kaizen situation." The way to best determine reality is to *observe* reality.

The rigor of PDCA and SDCA develop the practitioner's skills of analysis and assessment so that, among other things, waste can be identified, specifically the seven wastes. The best

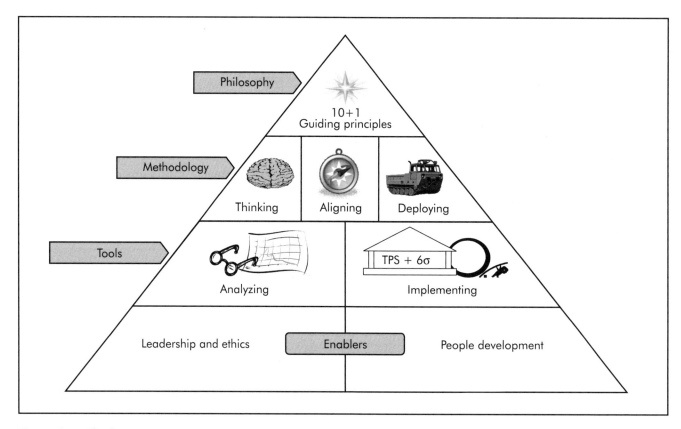

Figure 2-5. The kaizen system.

Kaizen Philosophy: 10+1 Principles

1. Think PDCA and SDCA, the basic scientific methods.
2. Go to the gemba; observe and document reality.
3. Ask "why?" five times to identify root causes.
4. Be dissatisfied with the status quo.
5. Kaizen what matters.
6. Have a bias for action.
7. Frequent, small incremental improvements drive big, sustainable improvements.
8. Be like MacGyver: use creativity before capital.
9. Kaizen is everyone's job.
10. No transformation without transformation leadership.

Do everything with humility and respect for the individual.

Figure 2-6. Kaizen philosophy.

way to identify waste is to observe that waste, firsthand. This is captured in *Principle 2—Go to the gemba; observe and document reality*. *Gemba* is Japanese for the "actual place" where the activities of a given process, both value-adding and non-value adding, occur. That place can be the office, lab, field, factory, mill, mine, hospital, or any other place in which material or information flows. More specifically, there are seven recognized flows, within which virtually all problems can be solved through observation and understanding:

1. person,
2. machine,
3. information,
4. engineering (and tools),
5. raw materials,
6. work in process (WIP), and
7. finished goods.

Principle 2 is founded upon the "three actual rule" in which the observer should:

1. go to the actual place where the process is performed,
2. talk to the actual people who are involved in the process, and
3. observe and document (chart) the actual process.

Genchi genbutsu, Japanese for "go look, go see," is a polite way of saying, "get off your butt, get out of the conference room, stop theorizing ("guessing"), and personally go observe reality and verify the data."

To understand what has been observed and documented, and discern the root causes of the waste, the one-word question, "Why?" is asked. More specifically, that question is asked at least five times... *Principle 3—Ask "why?" five times to identify root causes*. This practice, one in which most people were once experts (somewhere around the ages of three to five), is a lost art. And yet, it is the most powerful means of penetrating the veil of superficial "fixes" and excuses as to why things are the way they are, as well as shaking people out of their intellectual laziness or blindness.

Identifying waste is a wonderful thing, but it goes no further unless it is acknowledged so it can then be eliminated. In acknowledging the waste, another philosophical concept known as "the good" is encountered. Here a person moves from the realm of the intellect, the capacity that helped to identify the waste, and into the realm of the "will"—what is wanted or chosen. Choosing implies action (or in some cases, lack of action).

Human action is the natural consequence of a person's beliefs interacting with his or her desires: *Beliefs + Desires = Actions*.

Taking an immense philosophical shortcut, assume that lean zealots and wannabe zealots do not pursue kaizen for the heck of it. Rather, kaizen is undertaken for the foremost good of the company's stakeholders (customers, owners, employees, etc.) as measured by things like customer satisfaction, profitability, and employee morale. This is the desire to get better and do more than just survive.

Next comes the belief part. The desire must meet up with the intellect and its assent that the lean practitioner should give to reality. This reality drives the recognition that there are real competitors out there who are applying real pressure; there are real opportunities for improvement based upon real observation; and lean really works. Therefore, *Principle 4—Be dissatisfied with the status quo*, comes into practice. While that dissatisfaction is a noble and useful thing, there is greater leverage with proper focus. In other words, dissatisfaction can be generated by the comparison of the current state versus perfection (the waste-free condition)—truly an infinite field of opportunity. So, the focus needs to get a little more specific.

Invariably there are gaps in the measurable stakeholder satisfaction levels. This, when translated into the language of strategy or policy deployment and supplemented with the insight from value stream analysis and the related value stream improvement plans, should provide the company with specific break-through objectives, strategic initiatives, and deliverables. Kaizen activity must be focused, thus *Principle 5—Kaizen what matters*.

The combined dissatisfaction with the status quo and, more importantly, the existence of explicit performance gaps that are targeted for closure by a specific date should be unbearable enough to drive intense, aggressive action... not later, but now! This gives us *Principle 6—Have a bias for action*.

It is nearly impossible to have a bias for action and a simultaneous desire for the perfect solution. Taiichi Ohno preached against the ponderous efforts to design and implement elegantly engineered solutions. Quick and dirty is better than slow and fancy. This concept naturally leads to *Principle 7—Frequent, small incremental improvements drive big, sustainable improvements* (see Figure 2-7). This is certainly fundamental to daily kaizen, while kaizen events typically generate an intense and synergistic flurry of improvements that in turn yield more of a kaikaku effect.

The notion of small incremental improvement or elevation is synonymous with low- or no-cost activities. In contrast with innovation's typically substantial investment of time and money, kaizen is conducted with an emphasis on the quick engagement of pragmatic human effort and cleverness. This leads to *Principle 8—Be like MacGyver: use creativity before capital*. An equally flippant version of this principle is, "Reach for your brain before your wallet."

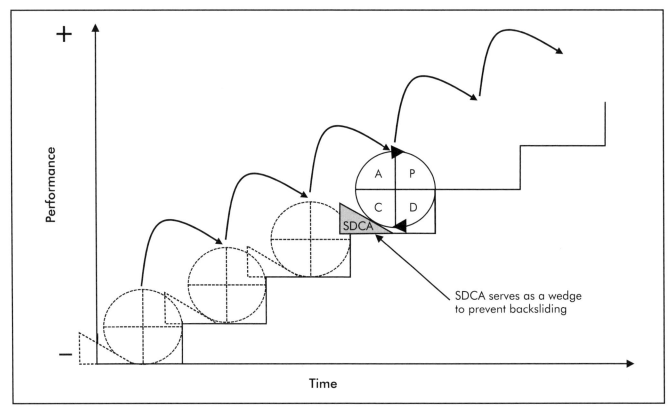

Figure 2-7. Small incremental improvements drive big results.

Kaizen is by no means elitist. While there are people in the workforce who desire no engagement other than to perform their day-to-day task, "maintenance" in the Imai diagram, they are few and far between. Rather, most people welcome the opportunity, really the responsibility, to improve their processes and those that are upstream or downstream. And, of course, as they say, "Many hands make light work." The workforce must be fully engaged to truly harness the engine for meaningful improvement. This leads to *Principle 9—Kaizen is everyone's job*.

While kaizen is everyone's job, it is largely wasted if leadership fails to lead. In the vacuum of vision, focus, consistency, and accountability bad things happen . . . and good things do not. Often the improvements that do get made will not be propagated and will certainly not be sustained. Leaders must personally live, facilitate, teach, and enforce all of the principles, and must do the same for the remaining elements of the kaizen system, its methodologies, tools, and cultural enablers.

Principle 10—No transformation without transformation leadership, expresses this simply. (This subject is discussed in detail in Chapter 3.)

Lastly is the *"Plus" Principle—Do everything with humility and respect for the individual*. This is central to both the kaizen system's cultural enablers and the lean business system. A person cannot truly listen or learn without humility. Further, kaizen is not done *to* people. It is done *with* people and for the purpose of achieving the balanced value objectives of all stakeholders. Of course, this does not mean that the gemba is a democracy. Lean leaders cannot abdicate their responsibility to do the right things for the enterprise, including making unpopular decisions and holding people accountable.

Methodology

Within the kaizen system, there are three categories of methods addressing the notions of thinking, aligning, and deploying. The core scientific methods of thinking (see Figure 2-8), specifically

> **"MacGyver"**
>
> "The series revolved around Angus MacGyver (known to his friends as MacGyver or 'Mac') who favors brain over brawn to solve desperate problems. MacGyver's main asset is his practical application of scientific knowledge and inventive use of common items—along with his ever-present Swiss Army Knife® and duct tape. The clever solutions MacGyver implemented to seemingly intractable problems—often in life-or-death situations requiring him to improvise complex devices in a matter of minutes—were a major attraction of the show, which was praised for generating interest in the applied sciences, and particularly engineering as well as providing entertaining story lines . . ." (Wikipedia 2009)

PDCA and SDCA, have been explored extensively thus far, and warrant no further discussion.

The category of aligning addresses the concept expressed in *Principle 5—Kaizen what matters*. This may appear obvious, but many people have affirmed by their actions that if the only tool available is a hammer, and there is nothing to tell anybody any different, everything looks like a nail. What usually follows is a lot of hammering and not a lot of meaningful progress.

The two primary means of effective focus and alignment for breakthrough performance are strategy deployment (also known as policy deployment or hoshin kanri) and value stream analysis or mapping with its resultant value stream improvement plan. However, depending upon the situation, for example a relatively non-repetitive process, it may make more sense to do process mapping and with that a process improvement plan.

Just-in-time (JIT) is one of the two major pillars of the Toyota Production System. One of the three elements of JIT is pull production. "Pull" reflects a dynamic in which the downstream operations, those that are closer to the actual customer, signal their requirements upstream. This ensures, among other things, that there is no overproduction—no overzealous banging of nails, especially the wrong ones!

In contrast, kaizen event alignment or focus (see Figure 2-9) is provided by the proxies of the customer and other stakeholders: the company's strategy deployment matrices, Gantt charts, and the various value stream or process improvement plans. These documents are typically summations of high- and detail-level analyses and have benefited from de-selection or filtering out of lower leverage opportunities. They ultimately identify specific, measurable, actionable, relevant, and time-bounded improvements that are necessary to move the process, value stream, and business from the current state to a desired future state. (Alignment tools are explored in Chapters 3 and 4.)

Improvement activities are essentially the deployment vehicles for kaizen. These vehicles (discussed in Chapter 4) usually take the form of kaizen events, projects, and what some call "just-do-its." "Just-do-its" are part "no brainer" and part "short putt" . . . provided that someone can provide the right energy and focus. Examples are things that require a policy change or edict from a responsible party, or a quick flurry of activity to create or enact a new procedure.

In the context of day-to-day (versus breakthrough) performance and the necessary daily kaizen, "what matters" is largely dictated and identified by: 1) the abnormalities and opportunities surfaced within the daily application of lean management systems and their gemba-based focus on process adherence and performance, and 2) empowered employees who, with their trained eyes for waste, are compelled to chip away at the barriers between the current state and the ideal state within a given process.

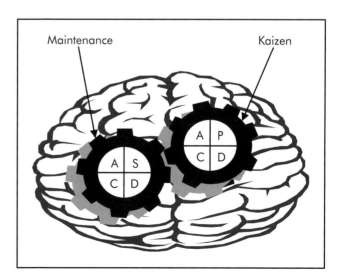

Figure 2-8. How to think: two core scientific methods.

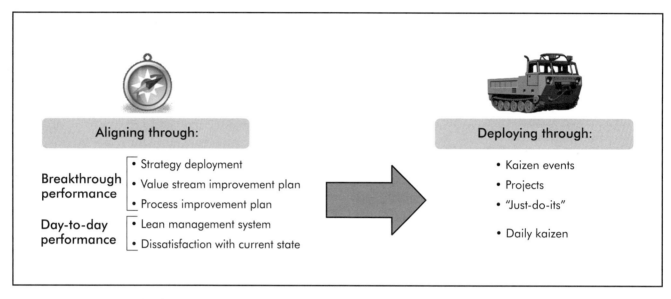

Figure 2-9. Kaizen methodology: alignment to deployment.

Tools

This kaizen system level is comprised of a host of lean and six sigma tools. The tools are separated into two general categories: 1) those for analytical purposes, and 2) those used for implementation purposes—the job of eliminating waste and reducing variation. (Both are reviewed in detail in Chapter 6.)

Six sigma's DMAIC methodology is the most obvious in its differentiation between analysis and implementation. As such, it is used here as an example.

- Analysis: *define*—make a project's purpose and scope explicit; *measure*—identify/quantify the location and source of the problem; *analyze*—identify the root cause(s) of the problem.
- Implementation: *improve*—implement improvements or countermeasures to address the root causes; *control*—evaluate the efficacy of the improvements and countermeasures; standardize to maintain the gains; and formulate steps for future improvement.

The separation between analysis and implementation is fairly clean until the *control* phase, within which both types of tools are employed. This "overlap" is due to the cycling inherent in PDCA. In lean, specifically in kaizen events, there is more dynamism in the use of the two tool categories. This is because of the short time frame, and therefore quicker PDCA cycling within kaizen events, which are typically 3 to 5 days, versus six sigma project durations of weeks and months.

So, What About Six Sigma?

Six sigma is a multi-faceted quality management system that, when properly applied, is synergistic with lean and kaizen. It is constitutive of the following elements or characteristics:

- Philosophy—it espouses a customer focus and process variation reduction through the identification and control of critical inputs.
- Methodology—define, measure, analyze, improve, and control (DMAIC) represents the primary six sigma methodology.
- Toolkit—six sigma applies an array of tools and techniques to facilitate the DMAIC methodology, including: cause-and-effect diagrams, failure mode and effects analysis, statistical process control, hypothesis testing, components of variation analysis, and designed experiments.
- Goal—"six sigma," from the perspective of process capability measurement (meaning six standard deviations between the process mean and either specification limit), reflects a process that yields no more than 3.4 defects per million opportunities.

The dynamism is due to a rapid "toggling" back and forth between measuring and improving and back again, and not due to a lack of tool distinction. This will become clearer as this book progresses into the detailed steps of kaizen events. In the meantime, Figure 2-10 provides some insight into the different tool categories.

Cultural Enablers

It takes years to nurture a continuous improvement culture and seemingly only weeks or months to destroy it. Consistent with that notion is the thought-provoking question that seeks an estimated continuous improvement or lean decay rate for a particular business or operation. It goes something like this, "If left alone without the typical 'extraordinary' intervention of a senior executive champion(s), outside consultants, or a lean customer who is requiring that you, the supplier undergo your own lean transformation, etc., how long would it take for your operation, value stream, or process to revert to its native, pre-lean launch state?" It is a nasty question, but it gets to the heart of the matter quickly.

The health and extent of an enterprise's cultural enablers are a key determinant in the answer to the decay question. This is why cultural enablers are the first of the four sections or dimensions of the Shingo Prize model. The others are: 2) continuous process improvement, 3) consistent lean enterprise culture, and 4) business results.

An organization has achieved a profound level of lean cultural maturity when it can be characterized as follows:

- "The improvement program becomes employee-driven rather than management driven;
- Lean concepts are applied in innovative ways outside the context in which they were conceived; and
- Improvements are made with the impact of all stakeholders in mind." (Shingo Prize model 2009)

The Shingo cultural enablers are founded on two very basic principles: humility and respect for the individual. These principles are in fact really virtues—operative good habits that are based upon right reasoning and the belief that lean is

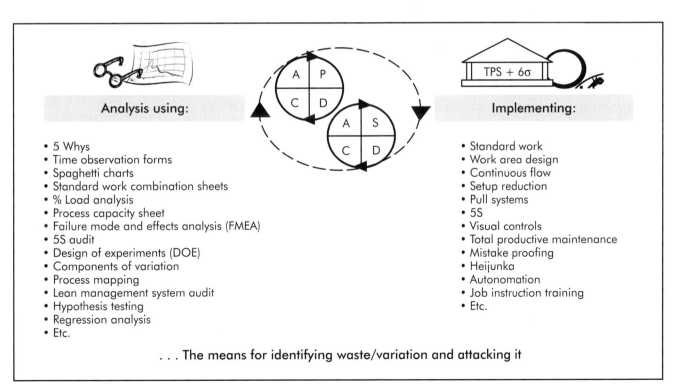

Figure 2-10. Examples of kaizen tools: analysis and implementation.

While the graphic will make a lot more sense after reading Chapter 3, this "no fish rotting from the head" text box will appear in each chapter. The primary purpose will be to share a relevant "lean leaders' 'must do/cannot fail' list." This list represents the things that all lean leaders MUST do for their enterprise to routinely conduct effective kaizen events, sustain the gains, and change the culture. Stay tuned.

the best way to do business. Perhaps it is based upon something that transcends even that.

Respect for the individual is a derivative of the virtue of justice—giving people their due. It is a mainstay of the Toyota Production System. Taiichi Ohno's early and simplistic schematic of the "Two Pillars of the Toyota Production System" reflected "respect for humanity" smack between the two pillars of just-in-time and automation with a human touch (*jidoka*). Within that figure is a notation that provides insight into respect for humanity (the term "humanity" perhaps more corporate in nature than "individual"). It reads, "Autonomy and a highly charged atmosphere in the workplace . . . Clarification of objectives. Joining forces to obtain the best knowledge and know-how . . . Desire to excel." (Japan Management Association 1989)

Respect includes basic considerations such as ensuring that people do not have to work any harder than they need to and that they are trained and encouraged to help improve the processes within the business through kaizen. Respect must be extended to other individuals, not only the workers, but other stakeholders, including customers, suppliers, and members of the community. Respect is "informed" and supported by humility, and vice versa.

The best students are humble. They are willing to listen and open to new ideas and concepts. Similarly, humble employees recognize that ideas can come from anywhere. Kaizen can be described as the flow of ideas—without humility that all-important flow is stemmed.

Consistent with the virtues of humility and respect for the individual, there are four primary cultural enablers:

1. *Leadership*—this is wholly in concert with guiding *Principle 10—No transformation without transformation leadership*. (See Chapter 3 for further discussion.)
2. *People development*—kaizen extends beyond processes and equipment to people, the only true appreciable asset, and encompasses behavioral and technical skills development. Fujio Cho, chairman of Toyota Motor Corporation, has referred to the Toyota Production System as the "thinking person's system." This thinking is reflective of employees learning and then applying the scientific methods of PDCA and SDCA.
3. *Empowerment*—when properly trained employees are encouraged and enabled to drive meaningful continuous improvement, they respond in awesome ways. This dynamic further embeds and propagates the lean culture within the organization.
4. *Environmental and safety systems*—these are reflective of a profound level of respect for both the individual and the community and concern for their well-being.

TOOL-, SYSTEM-, AND PRINCIPLE-DRIVEN KAIZEN

While somewhat different conceptually, both the Shingo Prize model and the Lean Certification Body of Knowledge, which is largely founded upon the Shingo model, recognize three different levels of lean transformation. The Shingo model employs the language of tool driven, system driven, and principle driven. Tool and system are reflective of different levels of the "know-how," whereas principle transcends the "how" to also include knowledge of the "why." Figure 2-11 reflects how the model differentiates the three levels in a kaizen context. It provides some very meaningful distinctions.

Unfortunately, many organizations, purposefully or unintentionally, seek to initiate their lean activity in a tool-driven manner. This is typically characterized by management planned kaizen events that are "focused" without the benefit of any substantial linkage to strategic direction or value stream imperatives. These events usually yield unsustainable and/or sub-optimized results.

System-driven kaizen, still event delivered, is management and engineering planned and explicitly linked to strategy and value stream improvement plans. Proper event planning, execution, and follow-through should generate substantial, sustainable results. While an event-only kaizen approach is an effective means of launching lean from the perspectives of early big wins, momentum, and organizational learning and engagement, it is not sufficient for a total lean transformation.

Principle-driven kaizen is the "gold standard" and is reflective of an organization that has built upon its successful kaizen event foundation and now enjoys a workforce that is engaged in both daily kaizen (predominately) and targeted kaizen events. The cultural and performance impact is deep and sustainable.

KAIZEN EVENT PULL

The application scope of kaizen events, in answer to *Principle 5—Kaizen what matters*, is largely determined by the previously described focus and alignment methodology and its elements of strategy deployment and value stream and process improvement planning. While it is necessary to focus the enterprise's limited resources on the vital few kaizen opportunities, this does not suggest that there must be absolute rigidity in the filtering of kaizen opportunities.

In fact, kaizen inherently has tremendous latitude and flexibility. As such, the "pull" can come

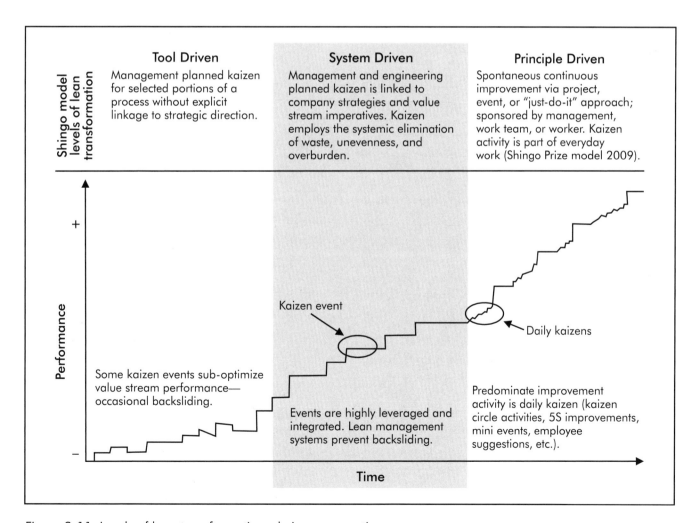

Figure 2-11. Levels of lean transformation—kaizen perspective.

from many sources. This dynamic is reflective of an enterprise that has transcended the purely systematic kaizen deployment approach and has begun to enter a deeper, more principle-driven kaizen culture. While principle-driven kaizen, like any kaizen, must be executed in accordance with standard work, it can and should be used to respond to the diverse challenges and opportunities identified in, by, or through:

- *The SDCA cycle.* By its nature, the SDCA cycle will identify new opportunities, even micro-opportunities to apply kaizen. For example, as a supervisor conducts her daily standard work, which includes review of the check processing standard work adherence, she notes that standard work-in-process levels are consistently being violated. In fact, the level tends to elevate to as much as four times the standard. This is indicative of at least three possible things (not necessarily mutually exclusive):
  - lack of adherence to standard work by the operators,
  - work content variation, and
  - less than optimal operator balance.

  Depending upon the answer, assisted by the application of the supervisor's 5-why analysis, kaizen may be required somewhere.

- *Performance gaps.* Performance gaps are characterized as significant shortfalls in actual versus targeted performance (for example, roll-throughput-yield, productivity, etc.). Not all gaps are anticipated (and coupled with the relevant gap closure plan) within the strategy deployment process and/or improvement planning. The lack of anticipation can be due to a variety of factors:
  - Simple oversight—it is possible that the performance gap was pre-existing, but it was lost or "de-selected" as the company sought to focus on the critical few gaps. No longer a gap amongst the trivial many, it now needs attention.
  - Gap creep—the performance gap, previously insignificant, has expanded such that it is now on the radar screen.
  - Dramatic changes in business dynamics—new, unforeseen forces, such as drastically increased material costs or a new, low-cost entrant to the market, have created substantial challenges.
  - Recovery from self-inflicted wounds—occasionally a plan is not well executed or was not very sound to begin with, or results in a product recall, higher production costs than anticipated for a newly introduced product, etc. In these situations, no one is going to ask whether the kaizen activity is in the strategy deployment matrix or value stream improvement plan. The problem needs to be addressed . . . now!

- *Opportunism.* This is the opportunistic application of kaizen in a target-rich area that was not necessarily "on the map" (part of strategy deployment deliverables, improvement plan action steps, etc.) but was later identified. The need may be precipitated by some drastic, unanticipated shift in market conditions. It is typified by the need for a quick hit to free-up

No Layoff Policy

One inviolable principle is that there should be no layoffs of permanent employees as a result of kaizen gains. This is both ethical and pragmatic. Ethical, because it is unconscionable to make gains through employee-driven improvements and then directly or indirectly terminate those same people or others within the same group, operation, or facility that is in the midst of a lean transformation. Pragmatic, because immediately after the first kaizen-driven layoffs, all volunteer kaizen participation will evaporate and forced kaizen participation will be unfruitful. Of course, the no layoff policy does not extend to situations in which business conditions have seriously deteriorated. As related by former Wiremold CEO and lean executive scion Art Byrne, there are at least "five lines of defense before showing people the door: 1) reduce overtime, 2) put the extra people on kaizens (to get future payback), 3) in-source some component [or services] from marginal suppliers [that the company] plans to drop anyway, 4) cut the work week across the board, and . . . 5) develop new product lines to grow the business." (Womack and Jones 1996)

capital, reduce lead time, or redeploy workers to some other value-adding activity.

- *Worker suggestions.* In a lean culture, employees will generate a number of suggestions. While many of these suggestions are of the "just-do-it" variety, there are some that may provide the impetus and framework for a kaizen. For example, a claims representative submitted a suggestion, which was to provide wider access to a system to determine check status on claim payments, thus eliminating hand-offs and customer wait time. This may present an opportunity for kaizen to more broadly address the empowerment of claims representatives and its potential to eliminate over-processing and waiting.

TWO BASIC LEVELS OF KAIZEN

Kaizen is multi-dimensional in nature. Event practitioners and participants alike will often refer to a number of different categories or types of kaizen, for example: point kaizen, paper kaizen, layout kaizen, process kaizen, machine kaizen, motion kaizen, setup reduction kaizen, office kaizen, flow kaizen, system kaizen, transaction kaizen, visual control kaizen, etc. Some of these descriptors simply repeat the lean tool name (for example, setup reduction), while others the venue or scope (for example, office), but only a few get at the different types or levels. It is important to understand the relationship between the different types because it provides insight into the hierarchy and sequence of design, scope selection, and implementation. Hiroyuki Hirano articulated this notion by differentiating between what he called point, line, plane, and cube improvements. These terms distinguish the progression of improvement scope and resultant lean functionality.

- *Point*—the point improvement is focused on a discrete area of a specific process, with the goal being aggressive elimination of the waste within the process.
- *Line*—the accumulation of point improvements among the "local" processes eventually enables the connection of same-stream processes into a "model line." A model line is a pragmatic lean implementation

Gemba Tales

Redeployment spurs personal growth. During a kaizen event, the team identified significant productivity opportunities. One of the operators within the target area understood this and the potential implications, including possible redeployment to another position. This no doubt was unsettling for a person who had performed the same job for 16 years. However, he had faith in management's "no layoff policy" and fully participated and supported the team's implementation of the various improvements. As expected, the resultant productivity improvements necessitated his redeployment to another position. Remarkably, after transferring, the operator realized that change and learning new skills was not as difficult as originally feared. In fact, it was an impetus for further personal growth. He decided to return to school and learn other skills, eventually leaving the company and beginning a career in the healthcare industry in a position that had greater earning potential and one which he found more personally rewarding.

Tale shared by Robert L. Klesczewski

strategy in which a prototype line or cell is developed using lean principles. This facilitates organizational learning by means of a real, small-scale lean deployment. The line is characterized by, among other things, the three elements of the first pillar of the Toyota Production System: just-in-time, namely continuous flow; pull scheduling; and presumably the orientation to and satisfaction of takt time. At this level, value begins to flow with minimal standard work-in-process inventory, through multi-process operations, and with the implementation of standard work, visual controls, etc.

- *Plane*—once the model line has been established, the lean technical and change management realities are better understood, and the various observers have had the opportunity to note that the improvement did not trigger the end of the world, it is time to replicate it to other lines. This represents a progression from one-dimensional improvement to two.

- *Cube*—no "line" is truly self-sufficient. It requires information flow, suppliers, product and process design, and so on. The third dimension recognizes the leverage of the full value stream, which leads to the ultimate inclusion of the full enterprise in the lean transformation.

The point-to-cube continuum concept (see Figure 2-12) is implicit within the two levels of flow kaizen (also known as system kaizen) and process kaizen. These levels are reflected in an adaptation of the original Imai kaizen diagram (see Figure 2-13) and provide insight into the associated primacy and focus.

The inherent logic is to start with the end in mind. This big picture, system view encompasses the value stream and its overall optimization. With flow kaizen, it is the responsibility of management to determine the future-state system and the means of getting there via the value stream improvement plan.

Once the big picture has been planned, then it is time to deep dive into the individual processes that, in combination, comprise the value stream. This sequence helps avoid the bane of many improvement activities—the blind optimization of a discrete process that unwittingly sub-optimizes the performance of the overall system. Process kaizen is the responsibility of the process owners and related subject matter experts who are tasked with driving the improvements specified by the value stream improvement plan. This rigor, however, does not preclude daily kaizen with its primarily point focus, addressing problems and opportunities highlighted by lean management systems or identified by engaged and empowered process stakeholders.

Flow kaizen is performed on the material and information flow of an overall value stream. The intent is to facilitate the flow of value by eliminating the waste (think of it as cholesterol) within the value stream. Beginning with the current state map of the value stream, practitioners seek to design a new and improved future state "system," to be achieved typically in the next 6, 12, or 18 months. The new system should reflect at least some of the following characteristics:

- The value stream satisfies customer requirements, including the rate of customer demand (takt time).
- A careful and explicit decision has been made to either satisfy demand through make-to-order or through a finished goods supermarket.
- Non-value-added steps are eliminated or at least significantly reduced.
- Flow of material and information has been achieved, wherever possible, based upon the characteristics of upstream and downstream processes (for example, process stability, cycle time, whether dedicated to the value stream or shared with other value streams, etc.). The pursuit of flow should follow the general hierarchy of:

  1. continuous flow,
  2. supermarket pull,
  3. sequential pull, or
  4. a hybrid of types 2 and 3.

- Information flow that specifies process requirements is sent as far downstream in the value stream as possible (pacemaker process).
- The demand, relative to volume, mix, and pace, is leveled at the pacemaker process.

The improvements necessary to transform the current state value stream reality into the targeted future state are delineated in the value stream improvement plan. This plan is constituted by process kaizens, projects, and "just-do-its." It must be executed with regard to approach, sequence, and timing so that the future state is realized by the targeted date.

THE LEAN BUSINESS MODEL: CONTEXT FOR ALL KAIZEN

The kaizen system must be holistic in nature to drive meaningful and sustainable improvements. Thus the true "home" or context of kaizen must be within the enterprise's own lean business system—essentially the framework for continuous improvement.

As part of an enterprise's initiation of a lean launch, or shortly thereafter, a tailored lean business system (LBS) model (see Figure 2-14) should

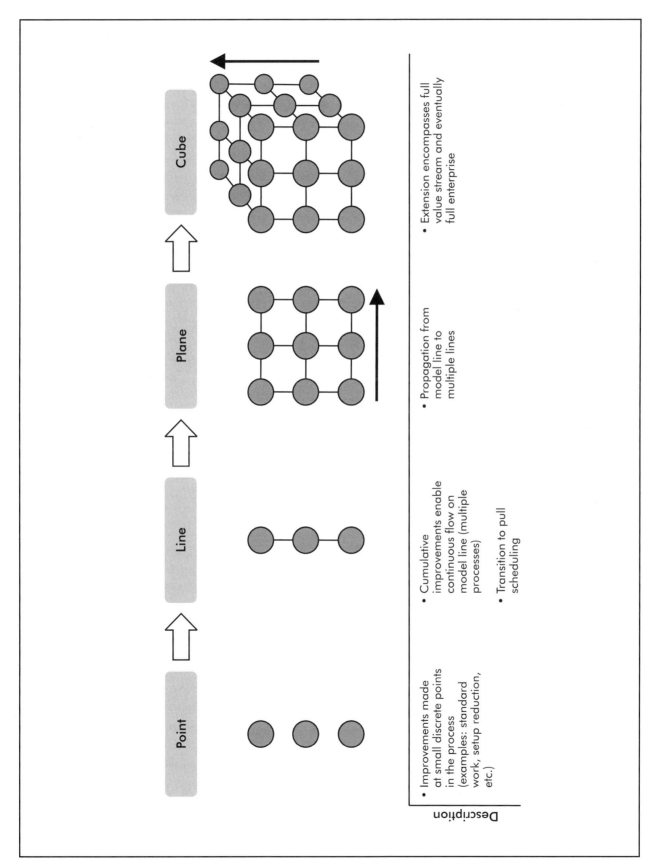

Figure 2-12. Cumulative improvements: from point to cube (Hirano 1990).

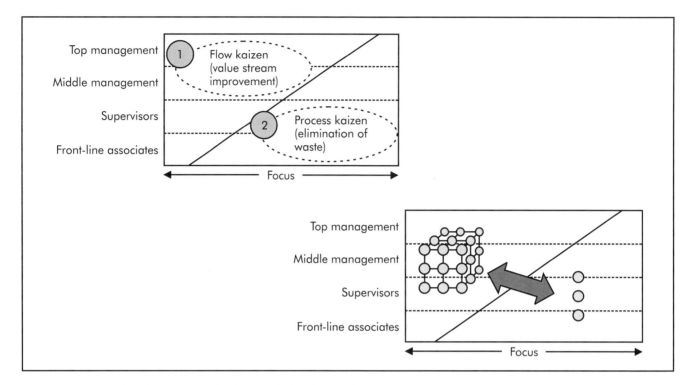

Figure 2-13. Two levels of kaizen (Rother and Shook 1998).

be articulated. This model is typically a variant of the Toyota Production System (TPS).

While it is easy to speak without thinking, a person must think before writing. The lean business system design process, one of the roles of the kaizen promotion office (KPO) (see Chapter 8), engages the mind and forces those who draft and adopt it to think long and hard about what is included and what is excluded. The model must convey a constancy of purpose, have a singular utility and integrity, and must be easily communicated. The LBS model should reflect, implicitly if not explicitly, the inherent relationships between cultural enablers, focus (customer, process, value stream, and enterprise), philosophy, tools, techniques, values and stakeholders, strategic direction and, ultimately, results.

Figure 2-14 captures the elements of a lean business system, which remain true to the core of TPS—JIT, jidoka, level loading (heijunka), standard work, and respect for the worker. This last notion serves as one of the central pillars. The worker, within the framework of the LBS, is the heart, head, and engine of kaizen. With the simultaneously scientific and creative dynamic of kaizen, people can accomplish amazing things.

The second central pillar, demand generation, is not a typical fixture within the LBS model. While many lean thinkers will occasionally mention a growth function, it rarely gets an explicit inclusion within the "house." Instead, the model is dominated by tools and techniques that effectively level and respond to customer demand and generate additional capacity. This represents a reaction to or readiness for pull, which if done with excellence, will satisfy existing customers and hopefully result in further business and securing new customers. However, pull response capability and ever-growing capacity must be balanced with pull creation, otherwise the model risks stagnation. Pull can be created through effective, lean-enabled processes that facilitate and drive:

- a deep understanding of markets and market opportunities,
- an agile response to market needs with appropriate products and services, and
- the crafting and delivery of a compelling message to the right potential customers.

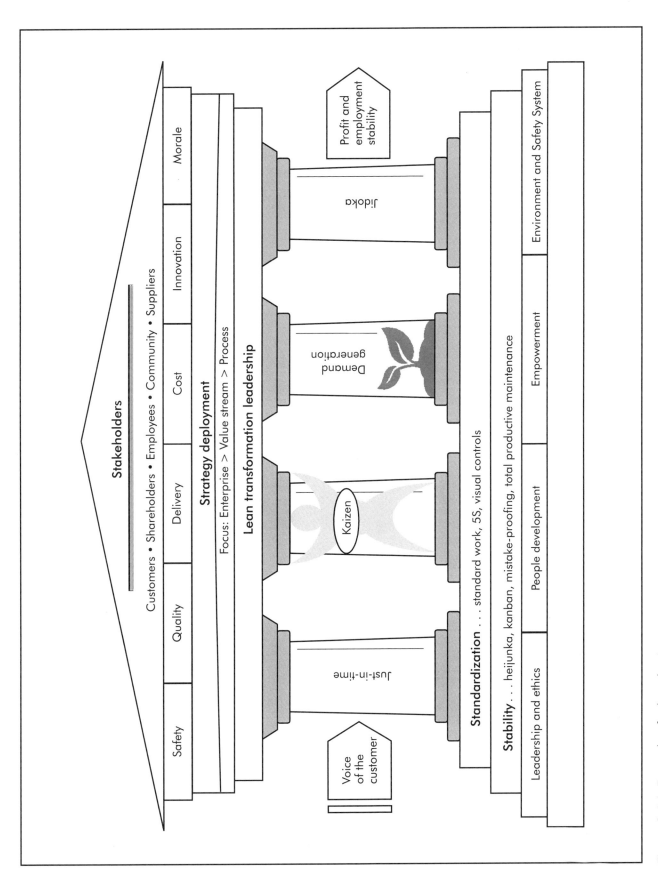

Figure 2-14. Example of a lean business system.

Of course, absent of effective leadership, much of the model is moot. (Chapter 3 addresses lean transformation leadership and the related topics of strategy deployment, value stream focus, and lean management systems.)

THE KAIZEN EVENT: RAPID IMPROVEMENT VEHICLE

The kaizen event is the rapid improvement deployment vehicle for incrementally and, often, in step-function manner, achieving strategic and value stream breakthrough results. It represents the aggressive application of PCDA/SDCA in which a cross-functional team, typically in the span of 3 to 5 days:

1. observes reality at the gemba (plant floor, office, lab, field, etc.) and identifies waste. (To the uninitiated this may seem daunting. While it does require hard work and focus, most of the "heavy lifting" can be accomplished using 10 basic waste identification tools and eight basic root cause analysis and supporting tools [see Chapter 6]. Because of the finite number of key tools, associates can develop a competency within a relatively short period of time.)
2. acknowledges the waste and takes meaningful action to eliminate it.
3. implements and verifies the new, least waste way.

As reflected in the different event name terminology, there is an emphasis on speed and profound change. Consider it the Special Forces version, albeit overt and not covert, of continuous improvement. It is characterized not only by speed and change, but also by leadership, teamwork, strategy, tactics, and tools. While the inherent risk from kaizen is quite different from that of a Special Forces operation, unsuccessful kaizen events still come at a cost. This cost is measured in time, money, morale, and lost opportunities, from both a business performance and human resource development perspective. Hence there is a real need for kaizen event standard work that encompasses event selection (strategy), planning, execution, and follow-through.

The Fieldbook is structured to walk the lean practitioner through the multi-phase kaizen event management approach as portrayed in Figure 2-15. Each phase is detailed, with its relevant standard work, in Chapters 4 through 7.

SUMMARY

- Kaizen is much more than an event; it is a philosophy, mindset and, for breakthrough performance, a most critical vehicle to achieve strategic imperatives and execute value stream/process improvement plans.
- Eliminating waste (muda) is the principle theme within kaizen. There are two types of waste:
  - Type 1 muda—wasteful activities that reasonably cannot be eliminated in the near term.
  - Type 2 muda—waste that is a prime candidate for quick elimination through kaizen.
- Muda is not the only focus. It is often accompanied by mura (unevenness) and muri (overburden or strain).
- The roots of kaizen are within Training Within Industry (TWI) and the basic scientific methods of SDCA and PDCA. Standards serve as the basis for improvement.
- The "kaizen system" is comprised of a methodology, tools, and certain cultural enablers, as well as a philosophy founded upon the following 10 + 1 principles:

A Kaizen Event by Any Other Name

There are different names for kaizen, but all have the same effect.

- Breakthrough event
- Continuous improvement event
- Kaizen blast
- Kaizen blitz
- Kaizen workshop
- Lean event
- Rapid continuous improvement event
- Rapid improvement event
- Rapid improvement workshop

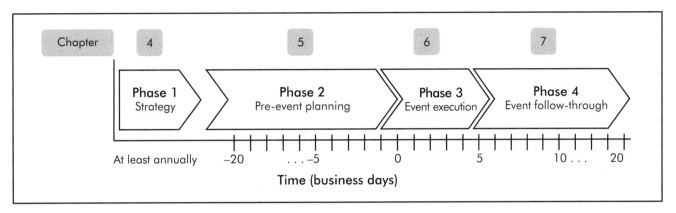

Figure 2-15. Kaizen event multi-phase approach—the x-axis reflects not only the sequence, but the minimum lead time for Phase 2, the recommended nominal cycle time for Phase 3, and the not-to-exceed lead time for Phase 4.

1. Think PDCA and SDCA, the basic scientific methods.
2. Go to the gemba; observe and document reality.
3. Ask "why" five times to identify root causes.
4. Be dissatisfied with the status quo.
5. Kaizen what matters.
6. Have a bias for action.
7. Frequent, small incremental improvements drive big, sustainable improvement.
8. Be like MacGyver, use creativity before capital.
9. Kaizen is everyone's job.
10. There is no transformation without transformation leadership.

Plus—do everything with humility and respect for the individual.

- The Shingo Prize model and the Lean Certification Body of Knowledge recognize three levels of lean transformation—tool, system, and principle driven. In the context of kaizen it is as follows:
  - tool driven—unfocused, management-planned kaizen events.
  - system driven—strategy deployment and improvement plan linked kaizen events that are management and engineering planned.
  - Principle driven—system-driven kaizen coupled with daily kaizen activities—spontaneous continuous improvement that is sponsored by management, work team, and worker.
- Kaizen events can be classified into two broad categories:
  1. Flow kaizens—these events, which are primarily a leadership function, focus on improving and optimizing the performance of the entire value stream.
  2. Process kaizens—when event based, the scope and timing are often established within a value stream improvement plan with focus on a specific process or sub-process within a target value stream. Process kaizen, in a mature lean organization, is driven significantly by daily kaizen activities.
- The lean business model should be the context for all kaizen.
- Kaizen event standard work is embodied in the multi-phase approach:
  - Phase 1—strategy (Chapter 4),
  - Phase 2—pre-event planning (Chapter 5),
  - Phase 3—event execution (Chapter 6), and
  - Phase 4—follow-through (Chapter 7).

REFERENCES

Businessdictionary.com. 2008. Accessed from Internet 4-18-08, Businessdictionary.com/definition/kaizen.html.

Cox, James F. III, Blackstone, John H., Jr., eds. *APICS Dictionary*, 9th Edition, 1998. Falls Church, VA: American Production and Inventory Control Society.

Deming, W. Edwards. 1986. *Out of the Crisis*. Cambridge, MA: Massachusetts Institute of Technology, p. 88.

Dictionary.com. 2009. Accessed from Internet 4-20-09, http://dictionary.reference.com/browse/kaizen.

Encarta World English Dictionary. 2009. Microsoft Corporation. Accessed from Internet 4-21-09, http://encarta.msn.com/dictionary_1861739570/kaizen.html.

Hirano, Hiroyuki. 1990. *JIT Implementation Manual: The Complete Guide to Just-In-Time Manufacturing*. Portland, OR: Productivity Press, pp. 31–34.

Huntzinger, Jim. 2008. *The Roots of Lean, Training Within Industry: The Origin of Japanese Management and Kaizen*. Accessed from Internet 12-3-08, www.lean.org/community/registered/articledocuments/roots%20of%20lean%20-%20TWI.pdf, p. 11, 14.

Imai, Masaaki. 1997. *Gemba Kaizen: A Commonsense, Low-Cost Approach to Management*. New York: McGraw-Hill, p. 3.

Japan Management Association. 1989. *Kanban Just-in-Time at Toyota: Management Begins at the Workplace*, Revised Edition. Portland, OR: Productivity Press, p. 25.

Lean Enterprise Institute. 2003. *Lean Lexicon: A Graphical Glossary for Lean Thinkers*. Brookline, MA: The Lean Enterprise Institute, Inc.

Rother, M. and Shook, J. 1998. *Learning to See: Value Stream Mapping to Add Value and Eliminate Muda*. Brookline, MA: Lean Enterprise Institute, Inc., p. 6.

Shingo Prize model. 2009. Accessed from Internet 4/20/09, http://www.shingoprize.org/Download/AwardInfo/BusinessPrize/ShingoPrizeModel.pdf, p. 4, 21.

Wikipedia. 2009. Accessed from Internet 4-15/09, http://en.wikipedia.org/wiki/Macgyver.

Womack, James P. and Jones, Daniel T. 1996. *Lean Thinking: Banish Waste and Create Wealth within Your Corporation*. New York: Simon Schuster, p. 140.

3

TRANSFORMATION LEADERSHIP

"The fish rots from the head down."
—Old proverb

LEADERSHIP TRUMPS ALL

Absent of effective leadership, the implementation techniques reflected in the following chapters is, by and large, useless (muda). Why? Revisiting the full results from the obstacles to lean implementation survey cited in Chapter 1, it is clear that leadership, or lack of it, is the primary driver. While shortfalls in implementation know-how were identified as obstacles 31% of the time, change management obstacles were overwhelmingly prevalent (Lean Enterprise Institute 2007).

- middle management resistance: 36.1%,
- employee resistance: 27.7%,
- supervisor resistance: 23%,
- lack of crisis: 17.7%,
- backsliding: 12.2% (this is also a technical issue),
- viewed as "flavor of the month": 8.8%, and
- failure to overcome opposition: 3.9%.

Perhaps unreported, due to the fact that the survey was taken of managers and executives, is senior management/executive resistance. Regardless, there is plenty of resistance to go around. And while resistance certainly ebbs as a lean transformation takes successful root, it will never disappear altogether. As allegedly stated by T. S. Eliot, "There are no causes that are permanently lost, because there are no causes that are permanently won."

But why such resistance? Ostensibly, it is because of the magnitude of the necessary change. The word "transformation," especially when mated to the word "lean" is deserving of the following definition: "In an organizational context, a process of profound and radical change that orients an organization in a new direction and takes it to an entirely different level of effectiveness. Unlike 'turnaround,' which implies incremental progress on the same plane, transformation implies a basic change of character with little

---

What Does Rotting Fish have to do with Transformation Leadership?

The phrase, "the fish rots from the head (down)" is rather peculiar, but everyone seems to know what it means—ineffective leadership necessarily drives ineffective enterprises. The universal understanding is evidenced by the fact that one proverb scholar identified more than 30 translations of this truism. And, it certainly predates lean. The first known English transcription is reflected in the 1674 publication, *An Account of the Voyage to New England*. (It is not known if that voyage had an inept captain or if there really were rotting fish on board!) (Merriam-Webster OnLine 2007)

> **Lean Leaders' Short "Must Do, Cannot Fail" List**
> - Drive change through action first.
> - Demand adherence to kaizen event standard work (it is a condition of employment).
> - Apply kaizen to high-leverage needs.
> - Make kaizen results stick by following the rigor of a lean management system.
> - Establish and nurture the KPO function.
> - Once a system-driven kaizen (event only) foundation is established within the organization, begin to deploy daily kaizen.
> - Follow a proven change management process.
> - Engage, unify, and if need be, force vertical and horizontal alignment of lean leaders.
> - Be emotionally intelligent; identify, use, understand and manage emotions.
> - Gain personal lean competency.
> - Always walk the walk, even when no one is watching!

at both the physical and virtual level, along with newly introduced terminology, technology, activity, rigor, accountability, and empowerment. Clearly, the change is daunting from an emotional and technical perspective. Lean transformation requires people at all levels to be adept at leading. To facilitate this modest exploration of transformation leadership, the following four areas will be reviewed:

1. technical scope (the hard "what"),
2. transformation leaders (the "who"),
3. emotional scope (the soft "what"), and
4. a transformation leadership model (the "how").

Reasonable questions at this point may be, "Why start with the technical?" and "Why not the soft stuff first?" Quite simply, technical change must precede cultural change. This most definitely does not discount or marginalize the emotional aspects of transformation leadership. People learn lean by doing and act themselves into a new way of thinking and doing (see Figure 3-1).

TECHNICAL SCOPE

or no resemblance to the past configuration or structure." (BusinessDictionary.com 2009)

The change required to effect a successful lean transformation is profound. Little is left untouched, including the very culture (beliefs, behaviors, and assumptions), strategic focus, organizational design, value streams and processes

The lean business model is the proper context for all kaizen. Therefore, it is decidely narrow to address leadership only within the scope of kaizen events; it is more appropriate to view it within the framework of the broader lean transformation and the kaizen system (refer to Figure 2-5). That said, the Fieldbook is foremost a prag-

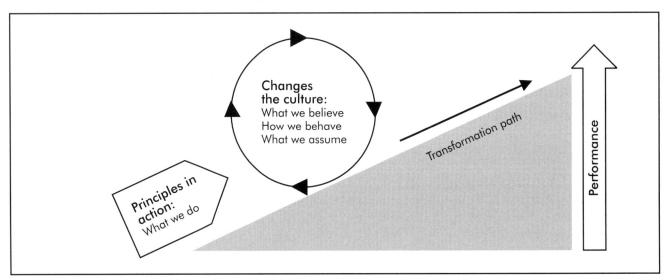

Figure 3-1. Principles in action change the culture.

matic reference for the kaizen practitioner . . . so it pursues a "middle scope" when addressing transformation leadership. This scope, portrayed in Figure 3-2, encompasses:

1a. *Kaizen event standard work.* Leadership is expected to consistently prescribe, apply, support, and improve kaizen event standard work. This must be done enterprise-wide and at all levels. In fact, senior executives must demand that this is done. The Fieldbook, by definition, should provide significant insight into kaizen event standard work.

Lean leader support is both material and emotional. Material support encompasses the necessary resources, such as available people to plan, lead, and staff the kaizen teams, facilitators and sensei, supplies, and training materials. Consistent with this requirement is the establishment and nurturing of a dedicated lean "function" or kaizen promotion office (KPO) function. (See Chapter 8.)

Leaders lend emotional support by means of encouragement, active participation, mentoring, and holding people accountable.

1b. *Daily kaizen.* System-driven kaizen (event only) is not sufficient for a total lean transformation. Rather, the enterprise must ultimately evolve to the next level, principle-driven kaizen, in which daily kaizen is regularly, pervasively, and predominately applied in conjunction with kaizen events. (See Appendix B.)

2. *The lean performance system.* Effective lean leaders deploy a lean performance system and consistently reinforce the rigor. The breakthrough alignment tools of strategy deployment and value stream improvement planning are combined with the day-to-day management of the lean management system. The breakthrough and day-to-day focuses employ PDCA and SDCA thinking. While the breakthrough side "feeds" the first phase of the multi-phase kaizen approach with high-leverage event scopes and targets, the day-to-day ensures sustainability of the kaizen event gains.

It is worthwhile to note that often strategy deployment is not used at the outset of a lean launch. This situation may continue for 12 months or longer. During this time, the organization is typically focused on a few critical value streams, seeking to drive substantial, quick wins via kaizen, and beginning to develop a lean organizational structure.

3. *The KPO function.* The KPO is a key resource for deploying kaizen event standard work throughout the organization. Its key result areas are: change management, company lean business system and curriculum development, people development, kaizen event management, daily kaizen deployment, kaizen promotion office management, KPO/lean deployment improvement, and KPO return on investment.

4. *Change management.* Not surprisingly, change management is about the process by which transformation leaders lead the change within their organization. From a technical perspective, there are best practices, which the lean leader would do well to follow. For example, change guru John P. Kotter outlines an eight-stage process of creating major change. These stages, related activities, and the Fieldbook relevant subjects are reflected in Table 3-1. Certainly, the

Gemba Tales

Lean fakers. The leaders of a particular business launched their lean transformation effort with impassioned speeches and written communications. Their primary impetus for the transformation, however, was purely opportunistic and short term. A significant customer was pursuing lean and this was seen as a great way to secure more business from them. During coaching sessions and kaizen events, the leaders could not be more positive. However, between events the commitment was non-existent. At the least sign of pain, the default was "business as usual." The commitment of those who had given their all during the kaizen event quickly waned. The charade finally came to an end after the company formalized a large contract with the target customer. Even leadership's superficial commitment dissolved—leaving those who had genuinely embraced lean feeling, to say the least, used.

Tale shared by Jerry C. Foster

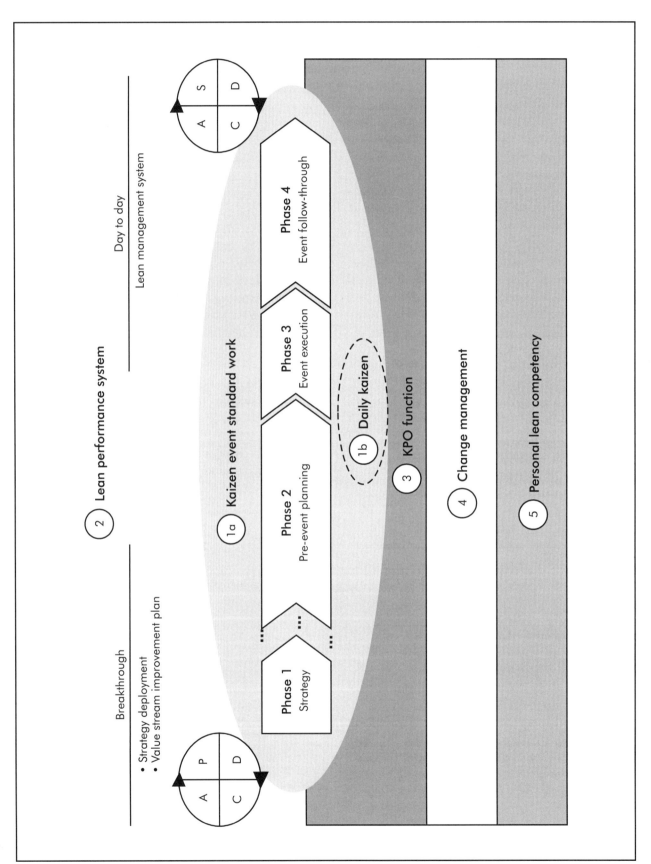

Figure 3-2. Fieldbook scope: lean transformation leadership.

Table 3-1. Change process (*Kotter 1996)

Stages*	Supporting Activities*	Fieldbook Relevance
1. Establishing a sense of urgency	• Understand market dynamics • Identify existing/potential crises/opportunities	Strategy formulation driving strategy deployment
2. Creating the guiding coalition	• Assemble team capable of leading change	Lean steering committee(s)
3. Developing a vision and a strategy	• Create vision to provide direction for change effort • Develop strategies to achieve vision	Strategy deployment, value stream analysis, kaizen
4. Communicating the change visions	• Consistently deploy all available communication vehicles to share vision • Guiding coalition role models expected behaviors	Kaizen-related communications
5. Empowering broad-based action	• Remove obstacles • Modify/remove systems/structures that undermine change vision • Encourage risk taking and out-of-box countermeasures	Kaizen, value stream analysis, strategy deployment
6. Generating short-term wins	• Plan/achieve visible performance improvements • Publicly recognize/reward people driving the wins	Kaizen and related communications
7. Consolidating gains and producing more change	• Expand change to address systems, structures, and policies that do not support the vision • Hire, promote, and develop people who can implement the change vision	Kaizen within context of value stream improvement plans, KPO function
8. Anchoring new approaches in the culture	• Create improved performance through desired behaviors, better leadership, and more effective management • Show causality between new behaviors and organizational success • Develop means to ensure leadership development and succession	Coaching within the kaizen process, lean management systems, kaizen-related communications and post-event audit

emotional dimension of change management should not be underestimated.

5. *Personal lean competency.* Lean leaders cannot delegate lean competency. They must develop it in the same manner as everyone else—through study and application. The requisite level of lean competency depends largely upon the leader's role and responsibility. Certainly, anyone within a leadership role should, at a minimum, get a sensei, read some good lean books, participate in one or more value stream analysis (flow kaizen), one or more process kaizen, and visit/benchmark some lean companies, without regard to industry. The key change agents and the senior managers, however, must do much more than the minimum. Indeed, lean thinking should become second nature to them. (See Chapter 8 for insight into lean competency development opportunities.)

THE LEAN PERFORMANCE SYSTEM

The "lean performance system" is a fancy name for the synergistic application of three elements:

1. strategy deployment,
2. value stream improvement plan management, and
3. a lean management system (LMS).

The lean performance system, governed by standard work, facilitates vertical and horizontal alignment of the strategic and value stream objectives throughout the enterprise. It drives proper lean behaviors, stimulates organizational learning, and instills management rigor. The lean performance system deploys relevant, high-leverage tools and techniques to achieve the objectives and then sustain and further improve performance. It is founded upon the plan-do-check-act (PDCA) and standardize-do-check-act (SDCA) cycles, applying timely, data-driven review so that abnormal conditions can be quickly identified and effective countermeasures implemented. The lean performance system dynamic is reflected in Figure 3-3.

Strategy Deployment

Strategy deployment is also known as policy deployment, strategic deployment, and the Japanese term, *hoshin kanri*. It is a process by which an enterprise translates its critical few strategic imperatives, typically in terms of 3- to 5-year measurable breakthrough objectives, "operationalizes" them, and deploys them across and down throughout the organization to the appropriate points of impact. Strategy deployment is purposefully distinct from the management of day-to-day performance, which is typically characterized by incremental improvement and tracked by things like key performance indicators. For example, a breakthrough objective may be "increase percentage of sales from new products from 5% to 25%" in three years, while a day-to-day relevant objective may be "increase order fill from 98% to 99%" this year. Breakthrough objectives are often necessarily cross-functional in nature, whereas key performance indicators often have limited cross-functional interface. The cross-functionality is evidenced by the broad assignment of deliverables, while vertical deployment is evidenced by the "cascading" of objectives throughout the organization until they get to the point of impact (in other words, to the specific person who is going to execute the specific deliverable).

Strategy deployment operationalizes the breakthrough objectives through a top-down, bottom-up collaborative process of defining and tracking. The process also includes a substantive amount of discussion/negotiation, known as "catchball," in which senior management and those who will own execution critically assess resource requirements and availability as well as the timing and magnitude of the desired performance improvements. The reality of finite resources requires the enterprise to focus on the vital few breakthrough objectives, typically three to five in total. The major elements of the strategy deployment process are summarized as follows.

- *Annual breakthrough objectives.* One immutable "rule" of lean is to constantly seek to reduce the management time frame (more frequent cycles around the PDCA and SDCA wheel) to drive better outputs. This rule is applied within strategy deployment, starting with the translation of the three- to five-year objectives into annual objectives. For example, the first-year percentage of sales from new products objective may be to move from 5% to 12%.

  Typically enterprise performance management systems are paced to an annual cadence. This may seem painfully slow, however, strategy deployment deliverables and related "checkpoint" reviews are monthly.

- *Strategic initiatives.* While the objectives are specific, measurable, and time-bounded, the strategic initiatives are directional in nature. They represent the overriding set of strategies that direct the organization toward achieving the breakthrough business objectives. The language of strategic objectives is "verb, noun," for example, "compress time to market." Consistent with the notion of focus, there should be only a handful of strategic initiatives.

- *Deliverables.* The deliverables, also called "targets and means," or simply "projects," represent the specific activities, each with measurable targets relative to magnitude and time frame that will drive the objective(s). Each deliverable should have a:

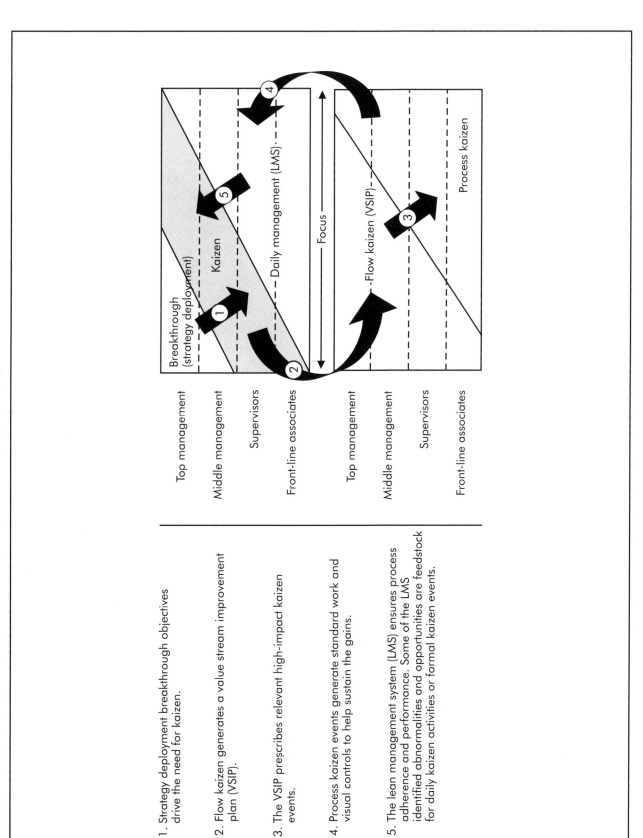

Figure 3-3. Lean performance system dynamics.

1. cause-and-effect relationship with one or more strategic initiatives,
2. specific owner who is responsible for its execution and the delivery of the expected impact, and
3. point-of-impact project plan (Gantt chart) that details time-phased action steps.

Monthly performance should be tracked against one or more key metrics to provide insight relative to deliverable execution and performance. These metrics are often tracked in a summary level "bowling chart," which compares the monthly target versus the actual performance and reflects a color to show status—usually green if the plan is met or exceeded, red if it is below plan.

- *X-matrix.* To facilitate communication, tracking, and review of the strategy deployment elements, teams at the relevant levels (corporate leadership, business units, value streams, etc.) will maintain a strategy deployment x-matrix. The x-matrix, also known as a hoshin matrix, reflects element linkage, ownership, and status as shown in Figure 3-4.
- *Checkpoint process.* Strategy deployment employs the classic PDCA cycle within a monthly checkpoint process. During this process, teams will, among other things, review the status of the items in the x-matrix, Gantt charts, and bowling charts, focusing on the red conditions (those not tracking to plan) and the sufficiency of/tracking to the deliverable owner's reported recovery plan. Necessary countermeasures are discussed or planned during the checkpoint.

Value Stream Improvement Plan Management

The output of value stream analysis for any given product or service family contains three elements:

1. a current state value stream map,
2. a future state value stream map, and
3 a value stream improvement plan (VSIP).

The VSIP (see Figure 3-5) represents a roadmap or game plan to transform the current state into the future state by the specified target date, typically through targeted kaizen events, "just-do-its," and projects. Strategy deployment and value stream improvement plans (discussed in Chapters 4 and 5) should be the principal drivers behind kaizen event selection relative to scope, timing, and sequence.

Lean leaders, many having directly participated in the value stream analysis, must apply the same level of PDCA rigor to VSIP review, execution, and adjustment as was done within the strategy deployment checkpoint process.

Gemba Tales

The fittest naturally de-select. Within the first hour of a three-day strategy deployment session, it became evident that the executive leadership team was focus-challenged. The preliminary three-year breakthrough objectives numbered eight, only three of which were captured in a single simple measurement. For the one-year objectives there was the same level of vagueness. The "first-pass" strategic initiatives were understandably loose and ill defined, yet also too numerous.

In a quick exercise of bold de-selection and rigor, eight three-year and eight one-year business objectives with over 20 strategic initiatives and over 70 deliverables were reduced to three breakthroughs for three years out. Commensurately, the first-year focus was clear, but no less audacious in stretch, with 10 strategic initiatives (still a lot) and 20 deliverables. These were revisited frequently throughout the year and re-graded when market conditions demanded. The PDCA portion of ongoing deployment tolerated no surprises, or at least none that lingered unaddressed, and ensured that the entire enterprise was working in a constant "get-to-green" mode. Year two of the strategy deployment process benefited from this solid foundation, enabling even more aggressive reach and broader inclusion of all functions and levels throughout enterprise.

Within four years, this sizable enterprise had worked its way to the top spot in its industry. Effective strategy deployment was not the only contributor to that achievement, but without it the company would still be playing second or third fiddle.

Tale shared by James J. Cutler

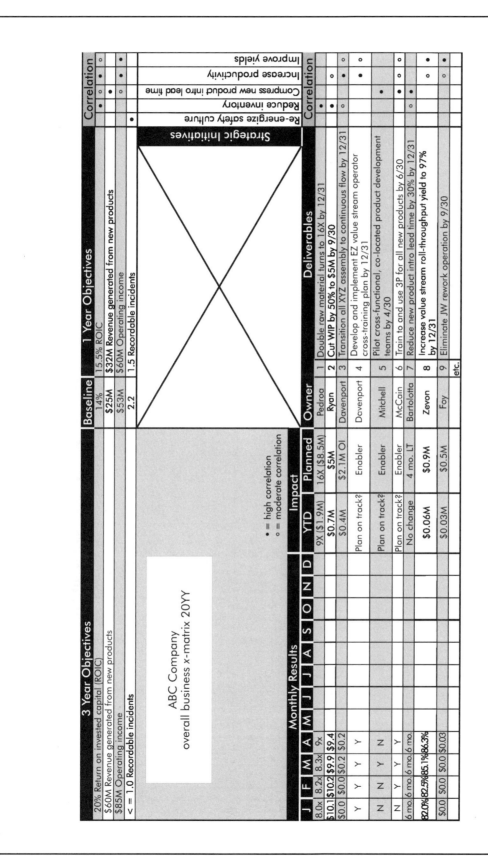

Figure 3-4. Example strategy deployment x-matrix. (For a blank form, see Appendix A.)

Value Stream Improvement Plan

Date prepared/revised: 12/7/XX
Product Family: Magdog
Value stream manager(s): Jack Martin
12/31/YY

Business-level objective: Establish lean value stream that compresses lead time to 8 days.

Value Stream				Measurable Goal	Summary Level Action Steps	Impact	Cost	Monthly Schedule...								Owner	Status			
Index	Loop Name		Objectives					Jan	Feb	Mar	Apr	May	Jun	Jul	Aug		25	50	75	100
1	Pacemaker	1.1	Communicate/ monitor plan and actual production sequence and timing.	1.1.1 Implement heijunka scheduling in mix area (tubs), and establish plan vs. actual by extrusion line by 3/31.	1.1.1.1 Conduct mini-kaizen to implement heijunka scheduling in mix area (tubs), and establish plan vs. actual by extrusion line by 3/31.	Med	Low			▓						Tinkers	25	50	75	100
1	Pacemaker	1.2	Improve customer lead time and optimize profitability.	1.2.1 Establish and articulate best practices for sequencing orders (include the following criteria: customer requested date, form, etc.) by 6/30.	1.2.1.1 Conduct sequencing process mapping meeting by 6/30.	Hi	Low						▓			Tinkers	25	50	75	100
1	Pacemaker	1.3	Improve extrusion process stability/ availability.	1.3.1 Reduce quality downtime by 50% by 6/30.	1.3.1.1 Conduct quality downtime kaizen event by 6/30.	Hi	Low						▓			Evers	25	50	75	100
1	Pacemaker				1.3.1.2 Rationalize in-process testing through business-level or local kaizen by 5/15.	Low	Low					▓				Chance	25	50	75	100
1	Pacemaker				1.3.1.3 Kaizen to optimize flow and standardize work in lab area by 8/15.	Med	Low								▓	Chance	25	50	75	100
1	Pacemaker			1.3.2 Reduce C/O downtime by 50% by 2/28.	1.3.2.1 Conduct C/O kaizen event by 1/26. Develop and implement standard work and plan vs. actual tracking.	Med	Low		░							Williams	25	50	75	100

Legend:

Impact	Hi	Med	Low
Cost	Low (Cost ≤ $10K)	Med ($10K < Cost ≤ $50K)	Hi (Cost > $50K)

Plan ▓   On schedule ░   Behind schedule ▤

Figure 3-5. Example value stream improvement plan (extract). (A form blank is available in Appendix A.)

The value stream manager, the line manager with responsibility (not necessarily full direct authority) for transforming the value stream, is the primary owner of the VSIP.

While value stream analysis in and of itself is not overly complex, achieving the future state is. There are five basic things, all within the lean leader's power, which must be avoided/addressed to have a fighting chance of success.

1. Current state and/or future state maps are not prepared or not prepared sufficiently. Unfortunately, value stream analysis is often done superficially or just plain incorrectly. Common issues include: mapping of multiple, unrelated product families within the same map, no time ladder, no data boxes, no roll-throughout line, and incomplete information flow.
2. The VSIP is not generated. Often people will generate maps and, after having identified the kaizen bursts, consider it done. They do not transfer the burst data to a VSIP and thus have no "plan."
3. The VSIP is incomplete. Some complete the maps and then populate the VSIP template with the required actions . . . and that is about it. Possibly the most egregious error of omission is the failure to identify a value stream manager. Clearly, this is not a VSIP "template" thing, but rather a lack of change management comprehension or lack of intent to ever make the future state a reality.

   Other overlooked elements include the timing and sequencing of VSIP action steps and the assignment of those actions to a responsible person. Ocassionally, value stream improvement plans are generated without sufficient information, such that many people, including those assigned the action steps, have neither a developed understanding of the approach nor the level of effort required.
4. The VSIP is flawed. Not all value stream analyses are consistent with lean thinking. As a result, the VSIP may reflect wrong or misguided action steps.
5. The VSIP is ignored. Many times the VSIP is developed and then never referenced again.

> **When Process Mapping Makes Sense**
>
> Much the same approach and rigor used within value stream analysis can be applied to macro processes. Process mapping and process improvement planning are an appropriate alternative to value stream mapping in certain circumstances—depending upon things like process complexity, repetitiveness, mix, process breadth (smaller), and the need for detailed understanding of the process. An example candidate for process mapping is financial reporting, with its often disparate number of reports. Process mapping drives the same outputs of value stream mapping:
>
> - current state process map,
> - future state process map, and
> - process improvement plan.
>
> It is often fairly easy to "back into" a value stream map for a macro process by taking the information, summarizing it into basic processes, identifying and quantifying the inventory throughout the stream, adding the time ladder, etc.

This is reflective of leaders who have a lack of understanding of or appreciation for value stream analysis and focus, an unwillingness to get into the details (for example, sequence, level of effort, required resources, and complexity), and the inability or lack of desire to hold people accountable.

Lean Management System

While strategy deployment and VSIP implementation are characterized by a definite rigor, they remain more oriented to breakthrough improvement and high-level PDCA application within a monthly checkpoint cadence. In contrast, the lean management system (LMS) is largely about day-to-day management, the sustainability of prior gains through process adherence, the identification of new opportunities for improvement (often daily kaizen), and development of the workforce (Figure 3-6). It embodies the near real-time application of SDCA and PDCA at multiple levels within the organization and is thus a profoundly effective mechanism for driving proper behavior

> **Project Management Skills**
>
> Lean leaders need basic project management competency. Specifically, they must be comfortable with developing and using planning tools that reflect: tasks, milestones, task assignment, timing, sequence, due dates, critical paths, dependencies, and level of effort. Similarly, lean leaders must be capable of managing project schedules and business risks (customer satisfaction, cost, technical, etc.) by applying basic PDCA/project review rigor, deploying recovery plans when needed, and holding people accountable.

and fostering lean thinking. In short, it is possibly the biggest contributor to cultural change and a requirement for sustaining any and all kaizen-driven improvement. Lean leaders ignore the LMS at their own peril!

The LMS is comprised of four elements (Mann 2005).

1. *Leader standard work.* Every worthy lean practitioner should have a solid understanding of standard work; the three elements of work sequence, standard work-in-process, and takt time, as well as the tools for developing and documenting it. Yet, the traditional notion is that standard work is principally for the "worker," a sometime euphimistic reference to everyone except supervisors, managers, and executives. This is flawed thinking.

Nearly everyone who has participated in a kaizen event has experienced the triumph of implementing improvements. Later, these same people, unless they permanently evacuated the gemba at the conclusion of the kaizen, have often experienced the post-kaizen hangover. This malady is characterized by a partial or full evaporation of the kaizen gains shortly after the kaizen. Sometimes this is due to two serious kaizen failure modes: 1) the team did not implement standard work, or 2) the standard work implemented was substantially inadequate. In the exercise of SDCA, these two conditions should be identified and the kaizen team "invited" back to finish the job.

SDCA application can also identify situations where there was a lack of worker adherence to standard work. Now, there may be a variety of reasons for the lack of adherence. The diligent use of five why analysis should identify the root causes. For example, training may have been insufficient, downstream materials or information may have been defective, equipment may have malfunctioned, or the worker may have decided not to follow the standard work.

The end result is that the "system" or "subsystem" that was implemented by the kaizen team, the new "normal condition," is not working as designed ("abnormal condition"), either consistently or intermittently. This is where leader standard work must be employed/expanded, preferably as part of or immediately after each kaizen event.

Precise, documented leader standard work requires team leads, supervisors, managers, etc., to audit the system at the gemba as aided by simple "drive-by visual controls." This is done at routine intervals so that it can be easily determined if the conditions, from both a process adherence and process performance perspective, are normal or abnormal. For example, the leader standard

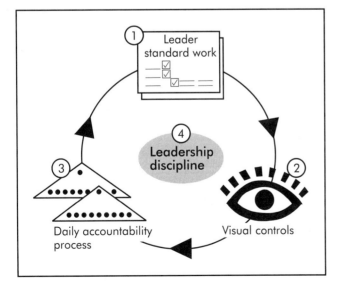

Figure 3-6. Lean management system.

work may call for the plant manager, once in the a.m. and once in the p.m., to review the status of the heijunka (load leveling) box and determine if it is loaded in accordance with standard work, that the cards are being pulled sequentially, that the cards are not being pulled prematurely, and that no cards remain in the box for previous time intervals (for example, at the plant manager's 10:15 a.m. review, there are no remaining cards within the box's 9:30–10:00 a.m. interval or prior intervals). In addition to audits, a team manager's leader standard work may reflect a daily (2:00 p.m.) requirement to review team member in-boxes and re-distribute/level the load among the entire team.

If an abnormality is detected, the good lean leader should note it within the leader standard work sheet, provide feedback to the operator and, in real-time ascertain the root cause and apply the necessary countermeasures. Sometimes the countermeasures are behavioral in nature (coaching); other times they are more technically oriented. Issues that are recurring and/or severe are typically elevated to the forum of the daily accountability process.

2. *Visual controls.* The execution of leader standard work is facilitated by visual controls. Well-developed visuals enable the observer, at a glance, to immediately determine whether a condition is normal or abnormal. Without nearly instant "drive-by" communication, audits become onerous, limiting their frequency and most likely the chances that they are done reliably. Further, simple visuals enable virtually anyone to discern the status of a process, thereby adding an additional level of rigor and attentiveness. No one wants their boss to identify the abnormal condition (and deploy an effective countermeasure) before they do themself. There are no secrets within a lean environment.

Of course, one abnormal condition that no visual control is immune to is the lack of maintenance of that visual control. Visual controls, by their nature, should be self-explaining and worker maintained. Sometimes, the worker does not maintain the visual (rendering it useless). The reason for this may be that the act of maintenance is complicated and/or time-consuming; the operator has not been sufficiently trained in its operation; or the operator may not be a fan of the transparency that the visual brings (relative to insight into process adherence or performance, or lack thereof). Accordingly, the leader standard work must require the leader to ensure that the visual is being properly maintained.

3. *Daily accountability process.* While the first two elements of the lean management system are largely about SDCA, the daily accountability process provides a routine forum and rigor to engage in PDCA. The process is effected through brief, standardized, stand-up, gemba-based, daily "tiered" meetings that focus on day-to-day performance, notable process adherence issues, and the necessary activities to improve the operation. The backdrop for the meeting is typically a set of simple, visual, and dynamic

Gemba Tales

Standard work; not just for "other people." During a lean management system kaizen, it became clear to the sensei that the plant manager did not fully comprehend the expectations related to the daily execution of leader standard work. Specifically, she did not think that it was something she had to do, even if it was for as little as 20% of her day. The sensei asked her whether she believed that virtually 100% of the operator's job on the floor should be governed by standard work. She answered in the affirmative. He then asked why, if it was expected that the workers adhere to standard work to ensure the least waste way, with the requisite quality and safety, would it not make sense for leaders to follow standard work to ensure that the "system" was working as designed and abnormal situations were quickly flagged and addressed? There was no reply . . .

boards (and flip charts) that facilitate quick review of the performance metrics, process adherence issues, suggestions, and improvement task assignment and statusing.

The tiered meetings, often called "sunrise meetings" start at the natural work team level (tier I) as led by the team leader. Depending upon the flatness of the organization, this meeting is followed by one, two, or three more tiers, usually concluding at the value stream or facility level. At each successive tier, the leaders of the prior tier participate, essentially rolling upward as the scope widens and expands beyond a tactical, 24-hour view.

As the tiered meetings progress upward, the cross-functionality of the team broadens with participation of not only line personnel, but direct support staff (for example, at the value stream level, this is the value stream leadership team). Consistent with this dynamic of increasing scope, the need for greater vertical and horizontal alignment, and increasing responsibilities, the meeting emphasis changes from tier to tier.

Tier I is focused primarily on sharing current day assignments, prior day performance, and related successes and issues, soliciting suggestions and reviewing brief, but key training points and other important notes. Subsequent tiers start out with an agenda similar (with an obviously expanded scope) to tier I. Each tier includes the quick review of the prior level, prior day leader standard work, assignment and statusing of team-member-owned countermeasures to address the issues and opportunities identified during the meeting and, depending upon the tiered level, a weekly review of kaizen newspaper items (see Chapters 6 and 7), and an at least monthly review of the VSIP (or process improvement plan) status.

4. *Leadership discipline.* Clearly, none of the first three LMS elements, either individually or collectively, will reach fruition without discipline. The requisite level of discipline can only eminate from dogged determination, self-management, and the willingness to hold people accountable from the top on down. This is especially the case during initial implementation of the lean management system when the naysayers will be in full force, protesting that "they don't have time to do this," "they already do this stuff," "it adds a bunch of muda," etc. Lean leaders overcome these obstacles and use the teachable moments, of which there will be many, to train and develop the team.

A well-developed lean management system is, by its very nature, easily audited. Lean leaders should make it a point to routinely audit the LMS. This can be done simply:

- Review a sample of completed leader standard work for completeness, recurring issues and problems, and evidence of good lean thinking in determining countermeasures.
- Walk the gemba with leader standard work in hand to determine its sufficiency and to observe, firsthand, the state of the gemba.
- Attend tiered meetings to determine the sufficiency of and adherence to the meeting agenda and the level of engagement, understanding, and lean thinking.
- Review the various supporting tiered meeting boards to assess the actionability, relevancy, timeliness of the performance measures, the type and status of assigned countermeasures, etc.

LEAN TRANSFORMATION LEADERS

The lean leader bench must be broad and deep to drive true enterprise-wide transformation. While a lean launch must be spearheaded by the chief executive, it cannot gain sufficient momentum, nor can it sustain anything meaningful, without the committed and effective enlistment of other lean leaders.

Leadership is a shared role. However, role clarity, as in any situation, is important. Here are brief descriptions of some of the most critical, albeit generic, leadership roles.

- The chief executive is the principal driver of the lean transformation. This senior leader

All is not Teachable

Stiles Associates, the industry's first lean-focused search firm, has been closely studying and validating the characteristics of effective lean leaders for nearly 20 years. They have identified a core set of measurable inherent traits that serve as a strong predictor of lean executive success. Their testing of over 2,000 people has yielded essentially "go/no go" limits on certain individual characteristics as well as ideal combinations. Some of these characteristics include:

- *nonconformity*. A lean leader cannot be wed to the way things have always been done. He must be ready and willing to explore different avenues to solve problems.
- *self-sufficiency*. The effective lean leader must be independent and self-motivated.
- *curiosity*. The five whys are not just a tool; they are a way of life.
- *impatient patience*. Successful leaders embody a strong sense of urgency that is tempered by reality.
- *imagination*. Starting with the end in mind, the lean leader must have a vision of what it will look like when it is done.

To these desired traits, Stiles has added a number of interviewing red flags to better identify a candidate's possible style and credibility gaps, which include:

- *frequently taking credit for all improvements*. Lean is team-based. Effective lean leaders are coaches that empower, teach, and bring out the best from their team.
- *inability to quantify results of past continuous improvement efforts*. This belies a lack of results and a plan, do, check, act headset.
- *overemphasis of certification(s)*. Lean is about transformation, not belts, certificates or initials.
- *lean understanding that is solely tool specific*. Lean is a holistic business system based on principles, systems, and tools, and it must be deployed that way.
- *terminal lean*. Real lean leaders know that transformation is a never-ending journey. Pretenders see it as a project or initiative that has a conclusion.

Based upon an August 12, 2009 interview with Matthew Ayers, VP executive search at Stiles Associates, LLC

sets the tone for the entire organization, making it crystal clear by example, teaching, and correction that certain behaviors, actions, and outputs are conditions of employment.
- Executives—because true lean transformation requires horizontal as well as vertical alignment, the full executive team must be committed. Too often lean is considered an "operations" thing and other functions are given a "free pass." A strong chief executive will require proper alignment of the various functions and disciplines.
- The steering committee and its cross-functional members are a prime example of Kotter's "guiding coalition." They address many of the programmatic and organizational dynamics within a lean launch. Their scope often encompasses transition issues relative to the progression from pilot to initial deployment to full-scale deployment, and personnel

Gemba Tales

Lean-er leadership. The prior regime had recently "transitioned" out and a new lean experienced vice president had taken over the reins. Former management's lean leadership was superficial at best and the kaizen efforts reflected the same. The executives rarely ventured beyond their offices and when there were kaizen report-outs they would come and listen—no questions, just listen. The new VP was different. At his first site report-out, one in which the kaizen team was explaining the newly implemented procurement visual management system that would significantly reduce material shortages, the new VP stood up and told the team to take him to the gemba and show him. He made sure that all 30+ attendees walked to the area. Once at the visual management board, the VP asked one of the procurement folks to explain how the system worked and how it would improve performance. Gulp!! With his action, the VP had unambiguously signaled that kaizen was important, leadership cared, and there would be a much higher level of rigor and accountability. Kaizen event effectiveness immediately improved from that point forward.

Tale shared by Bruce E. Thompson

policies surrounding worker redeployment, communication, and messaging. Steering committees are appropriate at both the corporate and business unit levels and may extend to facilities in which there is critical mass (for example, multiple value streams).

- The value stream leadership level represents a cross-functional team led by the value stream manager. Value stream leadership is responsible for driving day-to-day and break-through performance for a particular product or service family. The majority of improvements are directed by strategy deployment rigor, value stream improvement plans, and the daily kaizen opportunities identified within the lean management systems and by engaged and empowered workers who constantly compare the current state with an envisioned waste-free ideal state.

- Value stream managers are the line leaders with either direct or dotted-line reporting relationships with the functional support people (buyer/planners, manufacturing engineers, etc.) within product or service families. Together, they comprise the value stream leadership team.

- Functional managers/staff—while the value stream leadership team has primary responsibility for making it happen, there remains the need for staff functions at both the corporate and business unit levels. Staff roles are usually ones of subject matter experts, best practice developers, trainers, and support. Sometimes, as dictated by business need, they serve in integrator-type roles, such as project managers. Functional involvement within the value streams may be day-to-day if the value streams do not have enough critical mass to justify a full, cross-functional team. For example, a quality engineer may support two small value streams.

- Supervisors/team leaders represent those who typically lead natural work teams. Team member line of sight at this level is fairly parochial. Supervisors and team leaders play a critical role in communication and development. They facilitate team members' understanding of the need for change, their contributions, and roles within the bigger picture and empowerment within the context of continuous improvement and decision-making, all while ensuring process adherence and process performance accountability.

- Kaizen promotion officers are the key drivers behind the application and refinement of kaizen event standard work. They also serve as change agents and lean teachers (see Chapter 8).

- Kaizen event team leaders serve a crucial role throughout the pre-event planning, event execution, and event follow-through phases. Within the context of the kaizen standard work and related work strategies, they encourage and challenge team members to achieve the kaizen targets. They facilitate success by providing the team with the necessary resources and freedom (the "how") to deploy lean countermeasures.

Fortunately, most lean leaders are developed, not born. Providing that a person possesses the requisite foundational core competencies and technical aptitude, and operates within the framework of a lean performance system, much can be learned through study, application, and mentorship (see Figure 3-7).

EMOTIONAL SCOPE

"To be successful in implementing change in a company requires leaders who recognize the emotional impact significant change creates among organizational members and who understand how to minimize resistance to change." (Caruso and Wolfe* 2004)

When smart people behave unintelligently, it is often driven by the presence of strong emotions. Individual careers are frequently derailed due to a lack of attention to emotion and its related dynamics. Likewise, emotion often plays a major role in failed organizational change attempts.

Change management is most effective when the approach recognizes the feelings of the people who will be impacted—from the customer, to the front-line employee, all the way up to the executive boardroom. When circumstances are dire and a company is in danger of failing, initially the emotions may be fear, followed by hope. Executives who master turnaround situations know that, in floundering companies, complacency and false confidence can lead to corporate extinction. There must be a balance between the urgency of the impending change and the time available in which to bring it about. For example, the alternative approaches available to lean leaders span from a methodical, collaborative, coaching model all the way to a more autocratic approach, which quickly introduces specific technical changes that can drive the culture as people act their way into a new way of thinking and feeling. The best choice, however, is made through a combination of analytic and emotional intelligence.

For those readers who have experienced failures within their lean transformation, it is important to reflect on how emotions contributed to the lack of sustainable kaizen success. Emotional resistance at various leadership levels often has a dramatic impact on an organization's ability to implement a change initiative. Excellent ideas and innovative practices generated and implemented by committed teams can atrophy in the face of organizational resistance. Ironically, much of this resistance is generated by senior leaders who are reluctant to give up power, and middle- and lower-level managers who fear for their own positions.

Leaders and followers in organizations naturally form emotional responses to proposed organizational changes. Optimally, leaders and followers are expected to feel and behave in a supportive, enthusiastic, and engaged manner. However, fear, confusion, distrust, and sometimes anger are frequently encountered.

The kaizen event is a microcosm of a lean transformation, within which waste is public enemy number one. To facilitate the flow of value, unnecessary bureaucracy and functional silos must be removed. Often a value stream-based organization requires the elimination of entire management

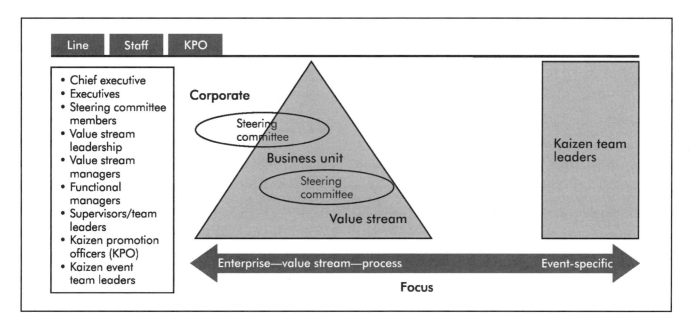

Figure 3-7. Lean leaders.

---

*Charles (Chuck) Wolfe is a professional colleague and friend of the author. He is a gifted leadership consultant and executive coach who has created innovative techniques that are critical for understanding leadership challenges. Chuck was asked to co-author this chapter to capitalize on his expertise and innovations in emotional intelligence (EI) and leadership. He has worked as a family and school counselor, psychotherapist, organizational behavior research associate at Harvard Business School, and director of leadership and organization development for Exxon and Hartford Insurance. In 1999, he was approached by Peter Salovey, provost of Yale University, and his colleagues, who asked for his assistance in finding ways to apply the newly discovered EI model and theory in organizational settings.

and supervisory levels and redeployment of those who previously staffed those positions. Similarly, many functions are decentralized to more effectively and directly support the value stream. Consistent with such changes to organizational design is the necessity for a different management style with less command and control behavior, and greater empowerment at lower levels.

Certainly, the magnitude and type of change that accompanies anything more than a superficial lean transformation will give cause for concern to an enterprise's employees. Not surprisingly, kaizen can also generate negative emotions in key individuals at all levels of an organization. This dynamic is explored further using the following quick case study of a kaizen event. It is actually a composite case from a number of companies put together to make the point.

The requisite pre-event preparation was done. At the event kick-off, a lower-level manager, filling in for the absent senior site executive who was tied up in "important meetings," gave a rather uninspiring and unconvincing speech regarding the importance of the event and management's commitment and expectations. By the time the team leader presented the pre-event area profile and target sheet, it was obvious that the level of team member engagement was flat at best.

The team gained perspective on the pre-kaizen situation through the application of time observations, spaghetti charts, and operator balance charts. Despite the fact that it was now evident that there was significant opportunity for improvement, the team did not exhibit any sense of urgency. Although people spoke, they seemed unwilling to get to specific countermeasures. Discussions were often circular and pointless, and frustration was building throughout the day. This situation skirted the radar at the first daily team leader meeting with management (the team had yet to hit the countermeasure wheel-spinning mode), but it certainly would not make it unscathed through the second.

The facilitator, a KPO who had recently been transferred to the troubled facility, tried a number of times to move the team along, resorting to several spicy exhortations with the team leader and team—all to no avail. Frustrated, he decided to seek out a steering committee member who was on site and with whom he had a relationship. He asked for some help to understand what was going on. He was told, "off the record," that senior leadership was engaged in a downsizing initiative due to the loss of a key account and he felt people were frightened at the prospects of reducing waste when it may have implications for their own jobs or those of their colleagues. Of course, any layoffs would first target temporary workers. While corporate had previously abided by the promise to not eliminate any jobs due to productivity improvements, sales downturns were understandably a different matter.

There were other issues as well. Many people who had participated in past kaizen events were frustrated due to the fact that the gains from previous kaizens were not sustained. This was largely because management failed to hold people accountable, especially when it came to adhering to the standard work implemented during the kaizen. Senior management's lack of rigor was evidence of not only poor leadership skills but also lack of commitment to the lean transformation. In fact, to stifle change, they would often reference the site VP's prior authorship of many of the current processes when he was the manufacturing director, "These processes and procedures were put into place by the existing site VP when he was in this role. Do you really think that your proposed changes are an improvement?!?"

The KPO thought about the situation and decided that evening to talk with the site VP. He did not share exactly what he had heard, but he suggested that the VP could give the kaizen team a big boost by coming down and talking about how he had helped to create the existing processes and procedures and what he hoped would be different as a result of this kaizen event. The KPO also shared that the event to date had not been as successful as he had hoped and wondered if the VP would also make his talk somewhat motivational if possible. While he was at it, the KPO respectfully "reminded" the VP how it would be appropriate for him to attend the daily team leader meetings.

The next morning, the VP came in and did an excellent job of discussing the way he had engaged in developing many of the tools in place today. He spoke about the excitement and enthusiasm he and a group of colleagues shared as they created and implemented innovative ideas that had enabled the firm to jump ahead of its competitors in a few key areas. He also said that his innovations were no longer advantageous and that the people in the room were largely picked because of their critical thinking skills, collaborative styles, and reputations inside the company. After he left the event momentum turned around and participants became more involved. Nevertheless, the downsizing issue weighed on everyone as well as the latent belief that the management team would, in the end, not sustain any gains.

There was still too much distrust and lack of confidence in the organization's follow-through for individuals to feel totally free to attack every area of potential waste and possible improvement. By the conclusion of the event there was very little implemented. The kaizen newspaper reflected a number of captivating new ideas, but as lean sensei often say, "The proof is in the pudding." In this situation, there was virtually no pudding! Distrust, apathy, and the past history of middle management's resistance to new ideas squelched implementation and marginalized the kaizen effort.

It is important to consider the role of emotions and how the lean leader can be smarter in responding to problems that are laden with emotion. For optimal effectiveness, the proactive use of the tools of emotional intelligence, and specifically the Emotion Roadmap™, is recommended. Consistent with this, the ability-based model of emotional intelligence (EI), conceived by Peter Salovey and John Mayer in 1990 and revised in 1997 (Mayer, Salovey, and Caruso 2008), will be explored. In addition, the Emotion Roadmap, which is the tool created by the author to help implement their pioneering model and theory, will be related back to the preceding case study.

The Emotional Intelligence Model

The four-branch model describes four ability areas that collectively describe emotional intelligence. This model is a refinement of the first formal models and measures of emotional intelligence (Mayer, Barsade, and Roberts 2008).

1. accurately identify emotions in yourself and others;
2. use emotions to facilitate thinking;
3. understand emotional meanings, how emotions work; and
4. manage emotions in yourself and emotional relations with others.

Using the Emotion Roadmap™

To apply the four-branch model in the workplace, a set of questions based on the four abilities (see Figure 3-8) is used to help individuals logically think through how best to handle any emotional situation that has high stakes and meaningful consequences. Certainly, a kaizen event meets these situational criteria.

The first ability, "identify," calls for recognition of the reality of current feelings. "Use" facilitates the creation of innovative and practical options that may be applied to affect the emotions the lean leader wants and needs to achieve the desired outcome. The idea that one emotion is better than other emotions for getting a specific task done has been proven in emotion research literature. A simple and well-known example of this is at a company's annual sales meeting where it is clearly important that sales people are enthusiastic, excited, and confident rather than anxious, doubting, fearful, and bored. Interestingly, it is rare that people spend any time thoughtfully matching emotions to specific tasks or events.

Think about a simple ongoing dynamic in every work setting such as an annual performance discussion. The event is often emotional to some degree and yet many organizations do not thoughtfully or proactively prepare managers to best address the emotional dynamics in these conversations. Think about the next performance review that you must conduct. How do you wish to feel in the beginning and how do you wish your employee to feel when you first start the conversation? What feelings would be ideal when you are in the middle of the discussion talking about needed improvements? How do you want both of

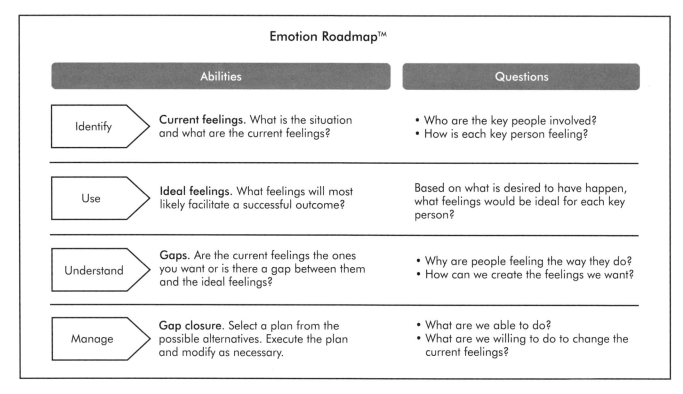

Figure 3-8. Emotion Roadmap™: Emotion-based planning and problem-solving process.

you to feel at the end of the discussion? Hopefully, reflecting on these questions helps you to realize how preparing for a potentially emotional situation at work can enhance your effectiveness.

It is obvious in the case study that kaizen event members need to feel secure. This requires that they trust that they will not be punished for suggesting, trying, and implementing different ideas and approaches. Also, they must be comfortable that their efforts will not result in the loss of employment. The difficulty in the case study is that the organizational history and actions do not engender trusting feelings. As suggested in the "use" section of the Emotion Roadmap, specifying the ideal feelings is critical for the event's success.

As apparent in the case study, there is often a gap between what people currently feel and the feelings that would be ideal for supporting an organizational change event. In considering that people are feeling distrustful, while the goal is for them to feel trusting, it is necessary to move to the "understand" section of the Emotion Roadmap. This will facilitate thought on how to generate feelings of trust. Understanding emotion requires having vocabulary to discuss feelings, knowing how feelings originate, and what makes them change. The objective is to change the existing feelings to ones that would be ideally suited for supporting and executing the kaizen event.

The "managing" step involves a review of the first three steps; that is, the identification of current feelings, thoughtful consideration of the ideal feelings, generation of alternatives to close the gap, and then selecting the best choice based on what the organization and key individuals are able and willing to do. The selection of the best-choice alternative is followed by executing and modifying as needed. The Emotion Roadmap can be used both reactively to an existing situation and proactively as a change management tool.

To further understand how the Emotion Roadmap can be a useful tool for lean leaders within the scope of a kaizen event, consider the case study from two perspectives. The first is the perspective of how the facilitator actually dealt with the situation and the roadmap. In this sense, the roadmap is reactively used as an emotion-based problem-solving tool. The KPO realized early on that the participants in the kaizen event were fearful and distrusting (Step 1, *identify*) when he wanted them to feel enthusiastic, engaged, and supportive (Step 2, *use*).

To understand how to change the feelings to the ones the KPO wanted, he sought out the steering committee member with whom he had a positive relationship. In their conversation he learned that there was a downsizing issue that was creating fear. Further, the past history of the organization had not been supportive of those who proposed improvements to existing processes or of improvements that had been implemented. Also, the current site VP had been the creator of many of the procedures within the kaizen target process. The KPO weighed his options and determined that the best alternative for generating the feelings he wanted the participants to experience (Step 3, *understand*) was to visit with the VP and get him involved. The VP was asked to assure the participants that he would welcome changes to "his" processes and procedures and to talk positively about the reasons they were selected for the event. The KPO did speak with the VP (Step 4, *managing*) and got him involved in a positive way. Ultimately, he got the kaizen team members to generate a number of useful ideas, a few of which were implemented.

The Emotion Roadmap as a Proactive Change Management Tool

If the Emotion Roadmap had been used proactively as a change management tool in the case study, the outcome would likely have been far more successful. The steps are the same, but they are done as part of, or prior to, the pre-event planning phase (Wolfe 2007).

In Step 1, the KPO might have asked the steering committee members to describe how the organization was doing. He would have asked if there were any other major initiatives that were taking place and he would have heard about the downsizing exercise that was supposed to be secret but was known by many in the company. He also would have asked about the history in the organization regarding the acceptance and prevalence of workforce-generated innovation. Assuming there was a level of candor with the senior leaders, he would have learned that people in the past were not encouraged to speak up and, in fact, in a few instances had been told directly not to criticize existing practice.

Having obtained this information (Step 2), the KPO would have been prepared to suggest some pre-event emotional preparation that would facilitate active participation and establish a safe brainstorming and "trystorming" environment. Indeed, the senior team needed to make people feel encouraged, supported, and enthusiastic about the process. Understanding (Step 3) what is involved in making a senior team turn a situation like this around requires some specific knowledge of individual leaders who are key to the event's success.

Every group, organization, and situation is unique to a degree. Nevertheless, the Emotion Roadmap can be used as a template to guide initiatives in a way that maximizes the opportunity for success. To apply the Emotion Roadmap as a proactive change management tool, here is how to proceed.

1. *Identify* existing emotions in the senior steering committee members. Most of these individuals arrived at a high organization level due to successful risk taking and innovation. As they have become accustomed to the prestige, power, and pay, they have sought to maintain status and keep a lower profile than what caused them to be noticed and promoted. They tend to want to avoid becoming enemies with other powerful people. Usually, they want to keep things the way they are because they have a clearly defined role and are part of the key group that has minimum vulnerability to the downsizings, right-sizings, re-engineering, etc., which takes place at lower levels. It is the rare executive who is totally self-confident in his or her ability to leave one organization to go to another and maintain or improve upon his or her substantial compensation package. This push toward conservatism comes from a fear of losing what has been gained, so there is a strong reluctance on the part of these individuals to support and encourage change.

In the case study, these fledgling executive lean leaders were only superficially and perhaps accidentally engaged. The only reason a lean implementation was considered was because the company had begun to lose market share, the overall economy was struggling, and no one seemed to have the answers as to how to improve. A number of the executives knew about lean, kaizen, and other approaches that had been used to help similar organizations improve. They made the decision to go ahead with an initial value stream analysis, which was followed by a handful of kaizen events . . . and then a re-evaluation.

2. *Using* emotions requires identifying the emotions necessary to create and sustain change. Executive lean leaders need to recognize that the status quo is no longer safe. They need to believe in themselves and that the abilities that originally caused them to be promoted still exist, and that strategic risk taking and innovation are their best chance for a positive future for their organization. Executives need to again rely on their capability for hard work, disciplined and focused planning, excellent critical thinking, and the ability to interact successfully with people above them. The ideal feelings are fear regarding maintaining the status quo, hope that they are the right group to turn things around, confidence that they can do the job, and commitment to getting it done.

3. *Understanding* emotions requires generating a set of alternatives that will significantly improve organizational performance, creating a sense of hope and engagement at all levels of the organization. Lean leaders must develop a collective vision of the future, one that is compelling and generates commitment to planning, implementation, and follow-through. In the case study, the executive team included several key individuals who were clearly the influential leaders. If the KPO had sought these people out and explained what needed to happen in a persuasive manner, the kaizen event could have been much more successful. The executives would have made sure that the downsizing initiative was clearly decoupled from the kaizen event. And, they would have delivered the message that the kaizen improvements would be sustained by management.

The KPO would need key executives from the steering committee along with the site VP to discuss how to create a shared vision that would encourage not only the kaizen event participants to become engaged and enthusiastic, but would also generate the desire of all those impacted to make the implementation and follow-through a resounding success. Thus it is also important to consider the composition of the steering committee. In all steering committees and leadership teams there are a few who are the key leaders who influence the thinking actions of the others:

- Who are they?
- What is their history?
- What are they feeling?
- What do they need to feel to become champions of change?

Creating a Vision

Whatever the strategy chosen for the work of transforming individuals and organizations, it needs to begin with an inspirational and compelling vision. A vision needs to be embraced at all levels to ensure its best chance of being achieved (Kouzes and Posner 2009):

"Yes, leaders must ask, 'What's new? What's next? What's better?' But they can't present answers that are only theirs. Constituents want visions of the future that reflect their own aspirations. They want to hear how their dreams will come true and their hopes will be fulfilled. We draw this conclusion from our most recent analysis of nearly one million responses to our leadership assessment, 'The Leadership Practices Inventory.' The data tell us that what leaders struggle with most is communicating an image of the future that draws others in—that speaks to what others see and feel . . . As counterintuitive as it may seem, then, the best way to lead people into the future is to connect with them deeply in the present. The only visions that take hold are shared visions . . ."

- Who is a champion or sponsor of the kaizen event?
- Who is neutral and how might that person be moved to become more supportive?
- Are any key leaders feeling negatively about the transformation initiative and, if so, is there a way to move them into a more neutral position?

4. *Managing* emotions begins with a review of the new alternatives with an eye to what the organization and its members are willing and able to do. For example, the lean leaders must discover, decide, commit, and execute the most successful initial value stream improvement plan. Once the plan is accomplished, other plans and actions are generated to build on the initial success.

In considering the launch of a lean implementation and with that, the necessary kaizen events, it is important to identify the key executives who are responsible for its success and how they are feeling at the moment. What feelings are ideal and is there a gap? If there is a significant gap that will negatively impact the lean implementation, what can be done to create the feelings required for it to be successful?

The proactive use of the Emotion Roadmap can prevent the fish from rotting from the head. It begins at the executive level, but does not end there. The roadmap should be progressively applied to the middle management group, supervisors, team leaders, and followers. Paying attention to the emotions in play will result in more successful events.

A TRANSFORMATION LEADERSHIP MODEL

After exploring the technical scope (the hard "what"), transformation leaders (the "who"), and emotional scope (the soft "what"), it is appropriate to discuss how those elements must come together to drive results. While obviously not comprehensive, Figure 3-9 reflects a simple model that attempts to capture the basic dynamics of transformation leadership. It is two-pronged in nature, recognizing the immediacy and impact of technical changes, and that employees will only take game-changing risk if there is a profound level of trust in leadership.

Humility and Respect for the Individual

The foundation of lean leadership must be humility and respect for the worker. Without this, no transformation is sustainable or worthy. Humility and respect for the worker in no way translates into a culture of feel-good superficiality or a welfare mentality. Rather, it recognizes the dignity of each and every person and his or her basic physical and emotional needs while at the same time affording a level of responsibility and accountability necessary to perform well and continuously develop.

Technical Change Drives Cultural Change

As discussed in the section on technical scope, technical change must take the lead in transforming the culture. Beliefs, behaviors, assumptions, and attitudes do not change through study, conferences, seminars, and training classes; they

Walk a mile in my shoes. We had finally convinced the president of the aircraft repair and overhaul business, "Bob" to participate on a kaizen team. The operators were doing a lot of walking to get special tools, up to 600 yards (549 m) one way. Since Bob did not get out on the floor much, the operators were all glad to see him and talk with him. Unfortunately, this worthy socializing was not conducive to good direct observation. He would constantly lose the person that he was supposed to be observing. After seeing the operator Bob was tasked to observe walk past me without Bob, I spied Bob chatting away with another operator. I "encouraged" him to catch up with his assigned person.

Bob responded, "I'll get his time when he gets back."

I reminded him that it wasn't just the time that was important, but that he personally needed to experience the effort that the operator was expending to get the job done. So I made him chase after the operator . . . almost running to catch-up.

After the observation, Bob said to me, "Man, that guy does a lot of walking. We need to do something about that!" Go figure.

Tale Shared by Craig Robbins

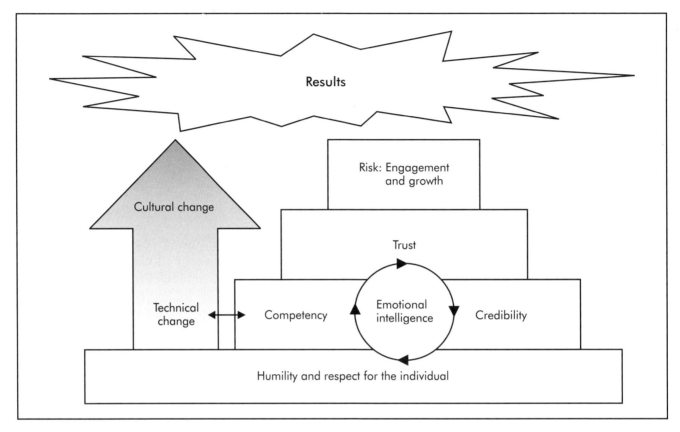

Figure 3-9. Transformation leadership model.

change through repeated action. This is not dissimilar to breaking unhealthy habits such as smoking or overeating.

The consistently repeated new lean actions and restraint from doing old non-lean things are undoubtedly "painful" in the beginning. Employees at all levels are required to learn and apply new tools and techniques. They must suppress old response mechanisms (for example, thinking there is no time to determine the root cause of those bad parts, just make more and cull out the bad so the shipment can be satisfied) and initially do things sometimes based largely on human faith. For example, they have been told that the new pull system will work, and even simulated it, but they have never really "seen" it work. Many of the new technical changes, such as leader standard work and visual controls, will introduce a whole new level of transparency and accountability. This will

"Shmed?" A KPO once remarked how another, newly minted KPO who had recently been moved from a plant manager role, called him to ask, "What is this 'shmed' stuff?" After some initial confusion (was it a new German pastry?), he finally determined that the new KPO was trying to ask about SMED or single-minute exchange of dies (a setup reduction approach). Unfortunately, the KPO who fielded the question was not discreet. He shared this funny story with many others in the organization. It did not help the organization's already unfavorable perception of the new KPO's lean competency. Not surprisingly, there was little "pull" for the new KPO's services from the value streams and her time in the role was short lived.

>
> **Gemba Tales**
>
> **Presidents' kaizen: A tale of two cultures.**
> As a young executive at what was once the world's largest car company, I experienced frustration as did many of my peers. The automaker, brimming with talent and resources, seemed disinterested in undertaking the change necessary to remain competitive. The young bucks knew that the enterprise needed to be profoundly transformed, so we each did whatever it took to dramatically improve the performance of our slice of the organization. However, the whole pie was immense . . . and the change effort was fragmented and inconsistent.
>
> Imagine the difference. During the second week with Danaher as a divisional VP of manufacturing and engineering, I was asked to participate in a "presidents' kaizen" along with all of the group executives, division presidents (hence the name) and vice presidents from across the corporation. Upon arriving midday Sunday at the host facility, we experienced "boot camp" training in just-in-time (JIT), Toyota Production System (TPS) style. Then, like special operations on a pre-dawn exercise, we were assigned to various teams early Monday morning to tackle significant challenges. With constant coaching and cajoling, we were learning and implementing TPS firsthand at the feet of some of Toyota's finest former executives. These senseis were the real deal: Taiichi Ohno's leadership team that drove the development and execution of TPS throughout Toyota.
>
> The dynamics were breathtaking; four to five teams comprised of various executives, all with serious competitive traits and none wanting to be outdone by the other. The team rosters included names like Byrne, Moffitt, Pentland, Koenigsaecker, and Consentino. Whole production lines were overhauled and relocated, creating one-piece flow where huge batches had been the norm. Paint lines were reconfigured for running color changes. Large machine tool changeovers were sliced from hours to minutes. Assembly lines were revamped for flow and customer-driven pull while liberating workers for other production needs, to the tune of often 20–50% productivity impact. All this was done in 5 days . . . by executives . . . in the presence of the workforce!
>
> This indoctrination provided a glimpse of what enabled Danaher to gain and maintain best-in-class performance. Its foundation was comprised of some simple, enduring lessons:
>
> - Lead by doing, not by deliberation.
> - Embody a burning bias for action.
> - Establish a constancy of purpose that is relentless.
> - Never cease the process of improvement.
> - Challenge the concept of "good enough" and replace it with "what's next?"
> - Constantly enhance the only asset that appreciates over time—the employees.
>
> *Tale shared by James J. Cutler*

be intimidating at first, but ultimately it will facilitate new behaviors and greater effectiveness.

Competency

The competency element of the transformation leadership model encompasses the previously defined technical items:

- kaizen event standard work,
- the lean performance system,
- the KPO function,
- change management, and
- personal lean competency.

Unfortunately, many leaders have not realized that competency cannot be delegated; it must be developed through study and application, often under the tutelage of a sensei.

Employees can sniff out an "empty suit" in a heartbeat. These under-equipped leaders who pretend that they either know what they do not, or maintain that it is not necessary for them to know, garner little respect and do not engender a meaningful level of trust. Such a leader may be a nice, well-intentioned person, but employees will only take risks if they have confidence that the leader knows what he is doing and will do it effectively.

Credibility

The flip-side of competency is character-driven—credibility. Despite the attempts of many leaders, neither involvement nor commitment can be delegated. Superficiality and hypocrisy are easily identified by employees. They look to their leaders in simple terms; they want them to "do what they say and say what they do." This extends to the rigorous application of the lean principles, systems, and tools. Lean leaders cannot, for example, bastardize the kaizen standard work. Cutting corners is typically counterproductive and models the wrong behavior for others within the organization.

Lean leadership requires an incredible force of the will to power through the initial barriers to change, including the whining, the complaining, the sabotage perpetrated by the lean antibodies, and the natural execution missteps. While initially virtually everyone will think the leaders are crazy, they must drive forward to establish credibility, demanding the implementation and use of things like daily accountability meetings, strategy deployment checkpoints, and gemba walks.

Emotional Intelligence

Emotional intelligence represents an awareness and (optimally) proactive response system to the emotions within the organization. It transcends and unifies virtually everything within the leadership model.

Trust and Risk

The enabling environment for healthy personal and group risk-taking is one of trust. This trust can only be achieved by means of consistent and convincing levels of leader competency and credibility.

SUMMARY

- For a successful lean transformation, leadership trumps all.
- Lean transformational leadership encompasses four areas:
  1. technical scope (the hard "what"),
  2. transformation leaders (the "who"),
  3. emotional scope (the soft "what"), and
  4. the transformational leadership model (the "how").
- Technical scope includes five elements of lean transformation leadership:

  1a. Kaizen event standard work—leadership is expected to consistently prescribe, apply, support, and improve kaizen event standard work.

  1b. Daily kaizen—once an enterprise has established a system kaizen (event only) foundation, then it must begin to rapidly deploy daily kaizen and evolve toward a principle-driven kaizen culture. (See Appendix B.)

  2. A lean performance system—this includes strategy deployment, value stream improvement plan management, and a lean management system.

  3. The KPO function—the kaizen promotion office is a key driver for applying and refining kaizen event standard work throughout the organization.

  4. Change management—this is the process by which leaders change their organization's culture and performance.

  5. Personal lean competency—these are the skills and experiences that lean leaders must possess, which cannot be delegated.

- Transformation leadership is a shared role spanning from the chief executive (the principal driver of lean transformation) to the kaizen event team leaders (who conduct the pre-event planning, event execution, and event follow-up). Each leadership level and role plays an important part in supporting the lean transformation. Fortunately, lean leadership is a skill that can be learned through study, application, and mentorship.

- The four-branch model of emotional intelligence comprises:
  1. accurately identifying emotions in yourself and others;
  2. using emotions to facilitate thinking;
  3. understanding emotional meanings, how emotions work; and
  4. managing emotions in yourself and emotional relations with others.

- The transformation leadership model captures the basic dynamics of transformational leadership. It recognizes the immediacy and impact of technical changes, and that employees will only take game-changing risk if there is a profound level of trust in leadership. This trust is built upon a foundation of humility and respect for the individual from lean leaders, as well as through leadership's competency, credibility, and emotional intelligence.

REFERENCES

BusinessDictionary.com. 2009. Fairfax, VA: WebFinance, Inc. Accessed from Internet, http://www.businessdictionary.com/definition/transformation.html, 1-6-09.

Caruso, D. and Wolfe, C. J. 2004. "Emotional Intelligence and Leadership Development" in *Leadership Development for Transforming Organizations*. Day, D. V., Zaccaro, S., and Halpin S., eds. Mahwah, NJ: Lawrence Erlbaum Associates, p. 237.

Kotter, John P. 1996. *Leading Change.* Boston, MA: Harvard Business School Press, p. 21.

Kouzes, J. M. and Posner, B.Z. 2009. "To Lead, Create a Shared Vision." *Harvard Business Review*, 87, January, pp. 20–21.

Lean Enterprise Institute. 2007. "Survey: Lean's No. 1 Obstacle can be Found in the Middle," in *Lean Directions*, September. Dearborn, MI: Society of Manufacturing Engineers, available from http://www.sme.org/cgi-bin/get-newsletter.pl?LEAN&20070910&4&. Accessed from the Internet 9/19/07. Note that the results add up to greater than 100%. Survey participants were encouraged to select all relevant barriers.

Mann, David. 2005. *Creating a Lean Culture: Tools to Sustain Lean Conversions*. New York: Productivity Press, pp. 25–26.

Mayer, J. D., Barsade, S. G., and Roberts, R. D. 2008. "Human Abilities: Emotional Intelligence." *Annual Review of Psychology*, 59, pp. 513–514.

Mayer, J. D., Salovey, P., and Caruso, D. R. 2008. "Emotional Intelligence: New Ability or Eclectic Traits?" *American Psychologist*, 63, pp. 503–517.

Merriam-Webster OnLine. 2007. "Words for the Wise Broadcast 10/19/07," Accessed 5/13/08.

Wolfe, Charles J. 2007. "The Practice of EI Coaching in Organizations: A Hands-on Guide to Successful Outcomes," in *Educating People to be Emotionally Intelligent*. Bar-On, R., Maree, K., and Elias, M., eds. Portsmouth, NH: Greenwood Publishing Group, pp. 176–178.

BIBLIOGRAPHY

Conner, Daryl. 1998. *Managing at the Speed of Change: How Resilient Managers Succeed and Prosper Where Others Fail*. New York: John Wiley & Sons.

PART II

STANDARD WORK: THE MULTI-PHASE APPROACH

4
STRATEGY—RIGHT WALL, RIGHT LADDER

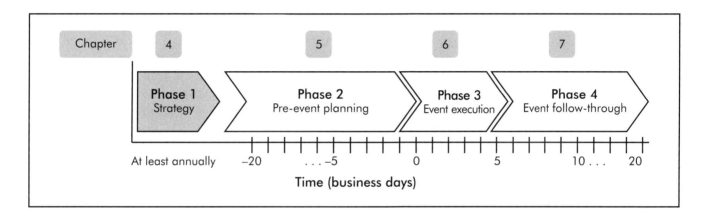

"If one does not know to which port one is sailing, no wind is favorable."
—Seneca

PHASE 1: STRATEGY

The first phase of the multi-phase kaizen event standard work is "strategy." This appears to be a generic and broad topic, especially in comparison to the following phases of pre-event planning, event execution, and event follow-through, all of which specify a work sequence and make use of things like checklists, forms, and meeting agendas. However, the strategy phase need not be esoteric. It may only seem that way because it is something that is routinely forgotten, ignored, or poorly done.

Within the context of continuous improvement, strategy represents the intersection between the alignment and deployment methodologies of the kaizen system introduced in Chapter 2. It is about figuring out how to achieve the enterprise's value objectives as efficiently and effectively as possible. Ultimately, there are two basic questions that need to be answered:

1. Is the scope and target high leverage (meaning high business impact with relatively low effort)?
2. Is a kaizen event the right deployment method?

In other words, the enterprise needs to first determine which wall(s) is worth scaling and, secondly, choose the best ladder to help scale that wall. Consider the alignment methodologies of strategy deployment and improvement

> **Real Nicknames for "Kaizens Without a Cause"**
> - Drive-by kaizen
> - Kaizen-in-a-box
> - Kamikaze kaizen
> - Popcorn kaizen

planning for value streams and key processes as the primary way to target the right wall, while deployment methodologies like kaizen events, projects, and "just-do-its" represent the basic ladder offerings.

The Right Wall

Without the requisite strategic linkage, kaizen Principle 5—"kaizen what matters" is violated ... often in a big way. The result is muda masquerading as kaizen, generating a lot of activity with little impact. Kaizen events require significant time, effort, and money. It is pure unadulterated waste when kaizens are not linked to what is important to the business, its customers, and other stakeholders.

> **Gemba Tales**
>
> **You want to what?** The Minnesota-based window manufacturer provided the sensei with the event scope sheets several days prior to the event (too late!). One of the sheets reflected a goal to improve the productivity of a *rework* process by 30%. The sensei thought that this must have been some sort of typographical error.
>
> During the Monday morning kick-off meeting, the team leaders reviewed their scope sheets with everyone and, as usual, the sensei followed this with a few comments and set some direction for the teams. The rework process team leader presented the scope, team, and targets (with the "error"). Next, the sensei asked if they were, indeed, serious about trying to improve the productivity of a process that was dedicated to waste. The team leader responded in the affirmative. The sensei countered that he would help them *eliminate* the process, but would not help improve it. The management team wanted to stay the course, whereupon the sensei said, "I hope you have a good week and we will see you on Friday. I cannot bring myself to go against the principles of my teachers and try to improve one of the seven wastes. Our job is to eliminate them!" True to his word, the sensei did not coach them. The team was a no-show for the report-out.
>
> *Tale shared by Craig Robbins*

A successful, high-leverage kaizen event invigorates team members. Conversely, a low-leverage event, whether successful or unsuccessful, is the proverbial eighth waste of a person. Team members invariably, and appropriately, identify whether the kaizen they are participating in is trivial ("busy work" or a "feel-good project"). Such situations rightly cause team members to question leadership's competency and credibility.

The Right Ladder

Assuming that the correct wall has been identified, the next task is to determine which deployment methodology is best. A kaizen event should not be the automatic default answer; instead the deployment methodology should be the least-waste way with requisite quality, delivery, etc. This means that, despite the power and flexibility of kaizen, there are times when a kaizen event does *not* make sense. See Figure 4-1 for a summary level flow chart reflecting decision logic for discerning whether a kaizen is appropriate ... or not.

Kaizen events should never be used to address "management" issues either from a behavioral or technical competency perspective. This is obviously inappropriate and ineffective. True lean leaders do not hide behind kaizen events and kaizen teams to deal with issues for which they are fully empowered, responsible, and accountable.

Similarly, in improvement opportunity situations where the current condition is well understood, the root causes have been substantially verified and valid countermeasures are identified, it is largely an execution decision and not a candidate for kaizen. When the effort and cost to execute is relatively small, for example, the elimination of a non-value-added report or changing a policy to empower certain employees to access a system to better perform their jobs, it is considered, as explained in Chapter 2, a "just do it." If the effort and/or cost to execute are significant, then it is often assigned, after a cost/benefit analysis, to one or more people as a project.

As discussed, kaizen events are team-based activities in which the members apply a scientific

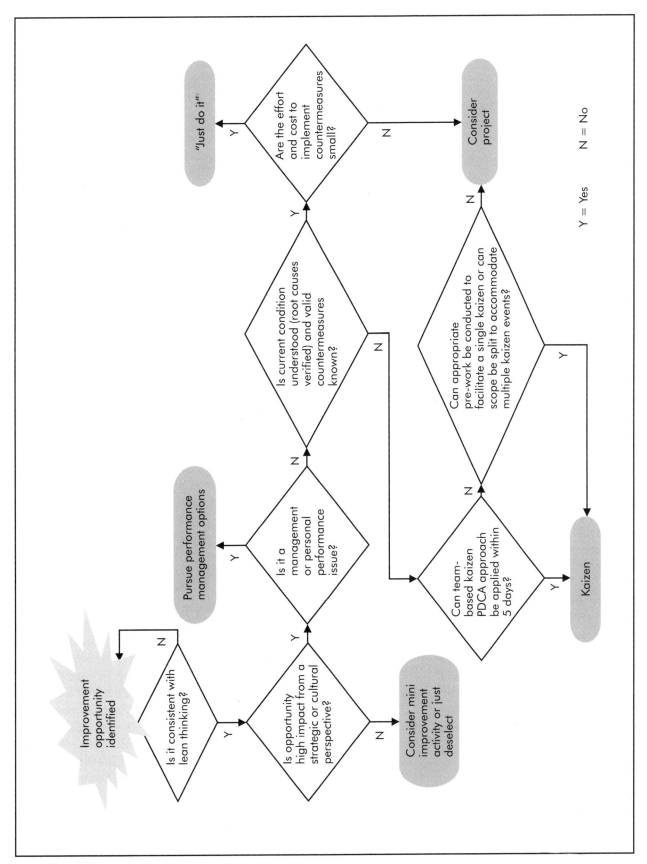

Figure 4-1. Decision logic used to determine if kaizen is the right tool.

Lean Leaders' Short "Must Do, Cannot Fail" List

- Use a kaizen event *only* if it is the right tool.
- Select kaizen events based upon an enterprise and value stream focus as dictated foremost by strategy deployment, value stream and critical process improvement plans, and/or A3s.
- Maintain and execute to a long-term kaizen schedule.
- Postpone or cancel events only in times of absolute crisis.

approach and rigor within a period of usually 3 to 5 days. Sometimes there are improvement opportunities that do not fit this model because of either scope or complexity. For example, a kaizen makes little sense for modifying software code or conducting an extensive designed experiment to select a specific machine setting for a given product.

While it is important to get it right when initially discerning whether an improvement opportunity is "kaizen-worthy," there is some risk mitigation. The rigor of the pre-event planning phase, when properly applied, *should* surface ill-proposed kaizen events and provide leadership with sufficient lead time to make the necessary adjustments. Of course, this is the waste of defects and requires the non-value-added activity of "rework."

Initial Kaizen Event Selection

The example lean business system (see Chapter 2) starts, from top-down, with stakeholder requirements, followed and supported by enterprise-wide strategy deployment, then value stream, and finally process focus. All of these elements are intimately linked and should almost always explicitly provide the purpose and focus behind each and every kaizen.

The sources of kaizen pull are characterized in Figure 4-2 relative to their planning or time horizons and their broadly categorized leverage or impact. The following sections further explain these sources of kaizen pull, their sequencing, and examples of their direct linkages to pre-event planning. For example, are events targeted first through strategy deployment, then by value stream mapping, or are there other paths? This dynamic is illustrated in Figure 4-3.

STRATEGY DEPLOYMENT

As previously noted, strategy deployment is typically not used at the first outset of a lean launch. This is driven by pragmatism; lean leaders want, as reflected in Kotter's seventh stage of the change process, to generate some quick short-term wins. Grand strategy can wait (often as much as a year or more) until there is some meaningful momentum. So while the examples herein start at strategy deployment, know that the lean launch usually begins with value stream analysis and a single value stream improvement plan.

In some situations, even the value stream analysis is deferred momentarily. This approach is usually only done for change management impact and features one (or more) quick kaizens on a "no-brainer" process that has a large amount of waste (visible at least to those with educated eyes for waste). Immediately after such a "splash" or jump-start type event, the resistant organization wakes up enough to be open minded for the next dose—value stream analysis, followed by a number of kaizen events as dictated by the value stream improvement plan.

Continuing from the strategy deployment overview in Chapter 3, the various *x*-matrices (see Figure 3-4 for example) reflecting the breakthrough objectives, strategic initiatives, and deliverables cascade throughout the organization (corporate to business unit to value stream, for example). This is done to ensure horizontal and vertical alignment. Ultimately, the cascading stops at the point of impact—an accountable person—actually a bunch of accountable people who each have one or more things that they must execute to satisfy the measurable and time-bounded deliverables. These deliverables will ultimately satisfy the breakthrough objectives.

At each point of impact there should be a Gantt chart detailing the things that need to get done

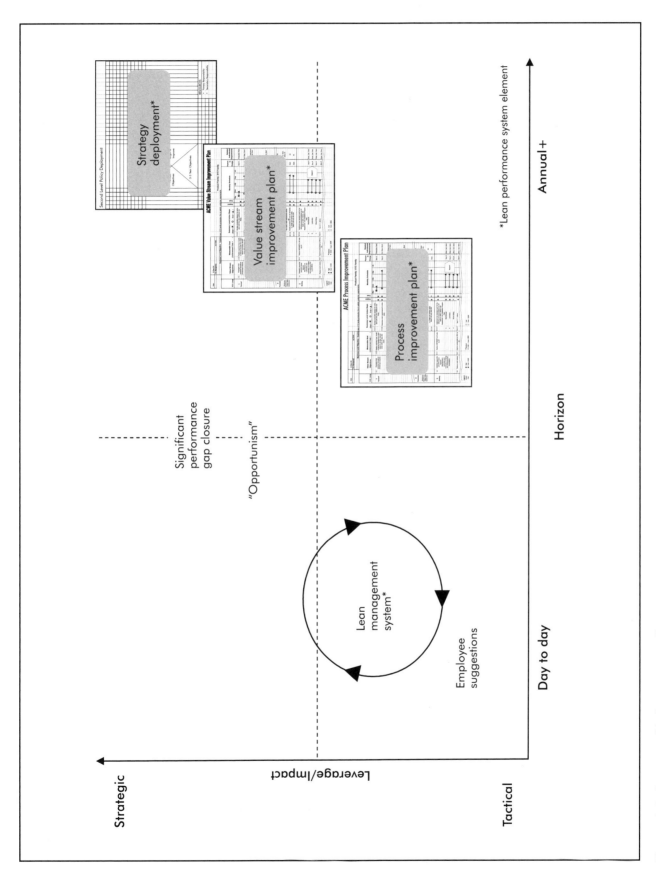

Figure 4-2. Sources of kaizen event pull.

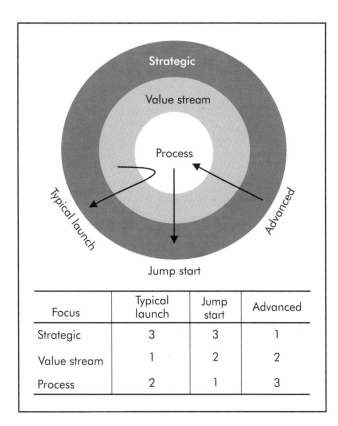

Figure 4-3. Kaizen event pull progression.

value streams of order-to-delivery and delivery-to-remittance.

Under-girding each of these value streams are the business support processes (finance, administration, human resources, etc.). Many of these processes, depending upon their criticality and the critical mass of the supported value streams, should be organizationally integrated within the very value stream that they support. Together, the value streams and the support processes represent the full-value delivery construct from which value stream analysis and process mapping generate:

- current state maps,
- future state maps, and
- improvement plans (see Figure 4-5), which are typically 6 to 12 months out and represent the roadmap for lean transformation through kaizen events, projects, and "just-do-its."

The improvement plans are time-phased plans for the implementation of the kaizen bursts. These bursts are visual and descriptive placeholders for the improvements that will have to take place for the future state map to become a reality. Each burst on the future state map, though it is sometimes difficult to physically fit them all in, should reflect three things:

1. directional objective with measurable goal, for example, reduce work content by 40%;
2. lean tool or otherwise that will be applied, for example, standard work kaizen; and
3. kaizen burst identification number, which helps to track the kaizen bursts as they are incorporated within the improvement plan.

Figure 4-6 shows an example of a process improvement plan. An example value stream improvement plan was reflected in Figure 3-5. In both examples, several of the "Summary Level Action Steps" have identified kaizen events. The corresponding "Measurable Goal" provides one or more specific, measurable, and time-bounded targets.

Other Sources of High-impact Kaizen

Kaizen would be an inflexible tool if its application was limited only to opportunities and

to satisfy the assigned deliverable. It is within this Gantt chart, and the corresponding "bowling chart," that the improvement activity scope, timing, and magnitude are articulated. The point of impact is where the pull for the kaizen is defined. See Figure 4-4 for an example.

Value Stream and Process Improvement Plans

The value stream represents all of the actions (value-added and non-value-added) required to deliver value, in whatever form (products or services), to the customer. At a macro level the value stream is constituted of three smaller streams of:

1. new product introduction,
2. inquiry-to-order (the sales conversion process), and
3. order-to-remittance.

These three value streams can be further dissected. For example, order-to-remittance may be decomposed into the more recognizable separate

DELIVERABLE: Compress lead time from personnel requisition to completion of new hire orientation from 132-day average to 60 days by 12/31/YY																	
LINK TO STRATEGY DEPLOYMENT Human resources-level 2 Level 1-compress lead time for new talent acquisition by 50%					OWNER: Jethro Tull						DATE (last update): Feb. 10, 20YY						
TEAM: Human resources (corporate and business-unit level)											NEXT REVIEW: Mar. 12, 20YY (monthly checkpoint)						
FACTORS TO CONSIDER: New hire process is poorly defined. Personnel requisitions typically do not reflect key information to conduct effective internal postings or outside searches. Multiple hand-offs, waiting, and rework occur throughout the process and there is possible over-processing with many multiple interview sessions for candidates. Long lead times are impacting execution within the three business units. Near-term needs are technical hires. Improvement activities must include multi-business unit, discipline, and level involvement as well as that of recent hires and possible outside recruiting firm(s) with whom we have long-standing relationships.																	
BOWLING CHART: Key measurement				Jumping-off Point	20YY YTD	J	F	M	A	M	J	J	A	S	O	N	D
Average lead time from personnel requisition to new hire orientation				132 days	Plan	132	132	130	130	125	120	110	100	90	80	70	60
					Actual	131	130										
#	Action Steps	Owner	Assist	Annual Impact ($)	Planned Date	J	F	M	A	M	J	J	A	S	O	N	D
1	Conduct cursory review of hiring data for last 12 months. Launch kaizen pre-event planning	Tull	Barre	Enabler	1/8	▓											
2	Conduct value stream analysis (scope to include: requirements definition, talent identification/recruiting, interview process, offer extension and orientation)	Tull	Barre	Enabler	2/4		▓	▓									
3	Execute "just do its"	Tull	Barre	Enabler	3/31			▓									
4	Conduct additional kaizen events as reflected in value stream improvement plan	Barre	Fuller	Enabler	5/31				▓	▓							
5	Pilot new processes within business unit	Barre	Jones	Enabler	7/31						▓	▓					
6	Deploy new processes corporate-wide	Barre	Tull	Enabler	9/30									▓			

Figure 4-4. Strategy deployment: Example bowling and Gantt chart. (For a blank form, see Appendix A.)

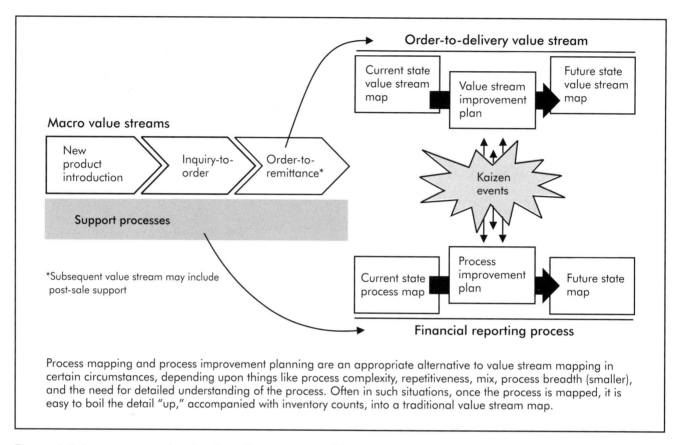

Figure 4-5. Improvement planning for value streams and key processes.

imperatives captured within the strategy deployment process and formal improvement planning. Business does not work that way; it is a little more dynamic and messy.

As reflected in Figure 4-2, there are other valid sources of kaizen event pull. The challenge, of course, is to ensure that these multiple sources do not encourage a free-for-all or the "kaizen-without-a-cause" phenomena. Here is a brief review of some of the other major sources of kaizen pull.

- *Significant gap closure.* The most important performance gaps are typically: 1) those that are allegedly being addressed within the strategy deployment process, but gap closure is not meeting plan (timing and/or magnitude), or 2) substantial shortfalls in key performance indicators like delivery to schedule and defects. These capture breakthrough and day-to-day performance, respectively.

The strategy deployment gaps are, by design, manifested in the monthly updated bowling charts (refer to Figure 4-4). When actual performance is below planned performance, this is considered a "red" condition as opposed to "green" (the monthly actual performance is shaded in the appropriate color). The strategy deployment deliverable owner must have a get-to-green action plan for such red items. Red items that persist for several months usually are escalated relative to detail and rigor of the recovery plan and its execution status. Similarly, key performance indicator tracking always should be compared versus a target. Significant shortfalls versus the target must also spur a recovery plan.

- *Opportunism.* The marketplace is a dynamic thing; technology changes, demand shifts, competitors enter and exit, material prices trend up and down, etc. Within this whirlwind

Process Improvement Plan

Corporate month-end financial closing 11/30/YY

Date prepared/revised: 6/15/YY
Process manager: Molly Michaels

Corporate-level objective: Establish a month-end financial closing and reporting process ≦ 7 workdays, by 11/30/YY

	Measurable Goal		Summary Level Action Steps	Impact	Cost	Apr	May	Jun	Jul	Aug	Sep	Oct	Nov	Owner	Status		
									Monthly Schedule						25	50	75 100
1	Existence of/adherence to macro month-end closing process standard work by May close	1.1	Conduct standard work kaizen to develop macro month-end closing process standard work and related milestones/close calendar by 4/25	Hi	Low									Evans	25	50	75 100
		1.2	Develop/implement lean management system to drive adherence to standard work, identify and respond to issues/opportunities, etc., by 5/15	Hi	Low									Fisk	25	50	75 100
2	Reduce correcting journal entries and related analysis by 50% on or before Aug. close	2.1	Conduct kaizen to develop process to drive accountability for data integrity throughout the organization by 6/25	Hi	Low									Evans	25	50	75 100
3	Achieve and maintain 100% Hyperion mapping accuracy to ensure accurate data by Sept. close	3.1	Develop change control process to ensure integrity of Hyperion data when changes occur by 7/1	Med	Low									Lynn	25	50	75 100
4	Transition from 5% to 15% value-added activity/analysis during month-end close by Nov. close	4.1	Eliminate tick and tie activity by loading report detail into Hyperion and spend time on value-added analytics by 8/31	Hi	Low									Lynn	25	50	75 100
		4.2	Identify/define value-added analytics by reviewing needs with stakeholders, best practices, etc., and incorporate into standard work by 9/30	Hi	Low									Rice			
5	Eliminate all manual data entry into spreadsheets (reduce work content and data entry errors) by Oct. close	5.1	Utilize data loaded into SAP and uploaded into Hyperion (after testing) by 10/15	Hi	Low									Lynn	25	50	75 100

Legend:

	Hi	Med	Low
Impact			
Cost	Low (< $10K)	Med (>$10K <$50K)	Hi (> $50K)

Plan ▮ On schedule ▮ Behind schedule ▮

Figure 4-6. Example process improvement plan (extract). (For a blank form, see Appendix A.)

someplace is usually unanticipated opportunity. For example, burgeoning demand may require the near immediate need for skilled workers. This may require significant work content reduction in one area so those workers can be redeployed into the high growth area.

- *Lean management system.* The rigor of the lean management system will quickly identify issues relative to the sufficiency of and compliance with standard work, process stability, and upside potential (see opportunism).
- *Employee suggestions.* The combination of engaged plan-do-check-act (PDCA) thinking workers, follow-through, and an un-bureaucratic suggestion system should generate a wealth of improvement ideas. Often these are of the "just-do-it" variety, but sometimes they are more substantial and may serve as feedstock for mini-kaizen activities or occasionally a full-blown kaizen event.

The A3 Report

There needs to be some rigor to help in the identification and framing of a candidate kaizen. The A3 report is a useful discipline for both problem-solving and proposing opportunities and the way to capitalize on them.

The A3 report, deriving its name from the international term for the paper size (297 mm × 420 mm, roughly equivalent to 11 in. × 17 in.), is a one-page storyboard designed to accomplish one of three basic purposes: 1) to facilitate problem-solving, 2) proposal formation, approval, and execution, and 3) the sharing of project status. Each purpose requires a different design, with the first fully exercising PDCA logic, the second, obviously focused only on "plan," and the third providing a check on the PDCA status. The brevity of all three A3 formats forces the author(s) to be extremely clear and concise, making the report easily understood by virtually any reader.

The first two types of A3 reports, problem-solving and proposal, are the most relevant to the discussion of kaizen event pull (see Figure 4-7). For example, the intent of the problem-solving A3 report clearly supports performance gap closure, while the proposal A3 report matches up with "opportunism." Of course, while the format is important, the thinking that must accompany its preparation is the most critical. This thinking, accompanied by that reflected in Figure 4-1, is what ensures a rational approach to identifying the specific need for a kaizen event(s) as opposed to other options.

Figure 4-8 shows an example of a problem-solving A3 report completed through the countermeasures portion, which highlights a kaizen event.

LONG-TERM SCHEDULING

Like most things, if kaizen event times are not blocked out on a schedule, they will not happen. Kaizen events are important, but they are not always perceived as urgent . . . at least until the culture changes. For this reason, most lean companies' KPO function maintains a 6- to 12-month rolling kaizen calendar, as depicted in Figure 4-9.

Scheduling is clearly done for the ultimate purpose of satisfying strategy deployment deliverables and executing improvement plans and the like, but it also has other practical purposes. Kaizen schedules are also formulated to facilitate proper internal resource allocation/rotation, secure external consultant time, spread kaizen activities among different locations and functions, communicate planned kaizen activities to employees, and provide adequate time for pre-planning.

Strategy deployment and improvement plans, with their longer planning horizons, are the real

Gemba Tales

Spread the wealth. One executive, determined to ensure that lean was not wrongly perceived as only a shop-floor "thing," made an edict that during every external multi-team kaizen event (those facilitated by an outside sensei), there would be at least one team dedicated to an "office event." This practice drilled into everyone that lean was enterprise-wide while also generating substantial and much needed improvements.

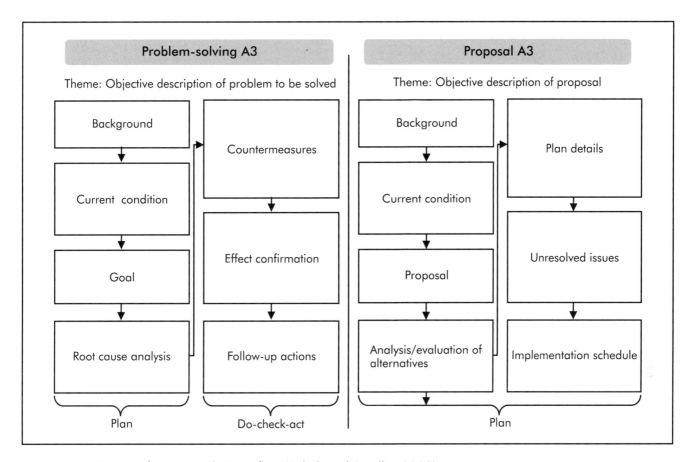

Figure 4-7. A3 report formats and PDCA flow (Sobek and Smalley 2008).

drivers of long-term kaizen event scheduling. These sources, along with well-developed A3 reports, specify event scope, targets, timing, and sequence (see Figure 4-10). The initial criteria for event selection and scheduling are threefold:

1. *Impact.* X-matrices, bowling charts, value stream and process improvement plans, and A3 reports target quantifiable improvements. These, coupled with less mathematically determinable outputs (for example, value stream improvements are system-based, so some improvements are "enablers" for other improvements), drive the impact or the leverage of a particular kaizen.
2. *Timing/sequence.* Gantt charts, improvement plans, and A3 reports articulate the timing and sequence for kaizen events.
3. *Resource availability.* While the aforementioned charts, plans, and reports identify the people responsible for the different improvement activities, they typically do not specify the total resource requirement for the kaizen events (team leader, team members, support personnel, etc.). Accordingly, when scheduling kaizen events, lean leaders need to ensure that they neither overload nor underload their resources.

Essentially, the object is to improve as fast as possible, without negatively impacting customer satisfaction, financial performance, employee quality of life, or the follow-through/sustainability of prior kaizen events. This means that, depending upon the critical mass, more than one 5-day event per month within a given value stream or macro process may be too much.

Of course, it must be remembered that improvement planning, usually done in the context of value stream mapping, is done at the 30,000-foot level. Therefore, the proposed scope, sequence,

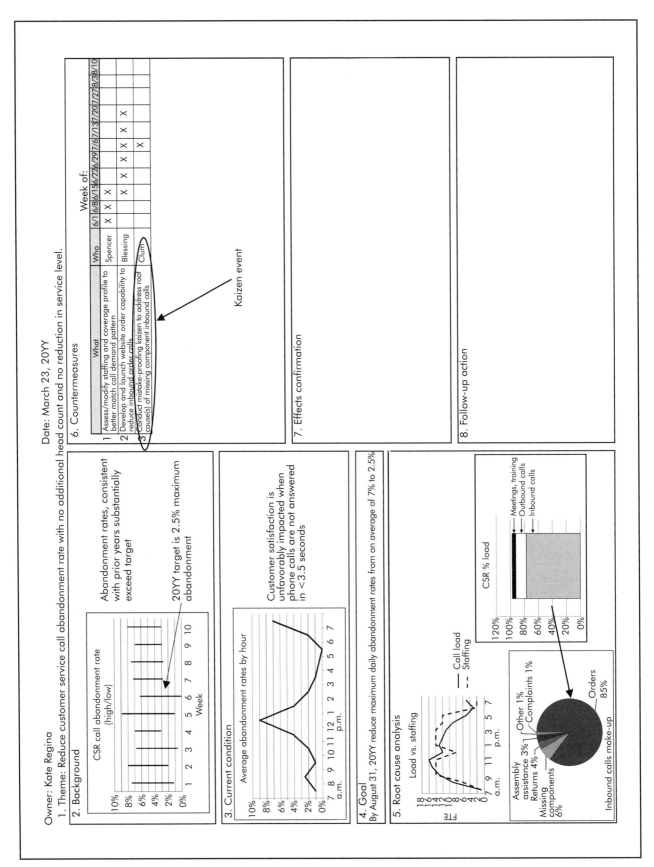

Figure 4-8. A3 kaizen pull. (For a blank form, see Appendix A.)

Master Kaizen Schedule

Date of last revision: Jan 20, 20YY   Business unit: Advanced Solutions   Scheduler: Beran

#	Date	Location	Event Number	Event Description	Sponsor	Linkage SD	Linkage VSIP	Linkage PIP	Linkage Other	Description/Reference	KPO	Internal	External	Consultant (if Applicable)	Event Select/Definition (1)	Communication (2)	Pre-work (3)	Logistics (4)
1	1/15-1/19	Avon	AV-#01JanYY	Setup reduction 40 mm extruder	Ortiz		X			Engineered products VSIP 2.3	O'Connor		X	Shingo	■	■	■	■
2	1/15-1/19	Avon	AV-#02JanYY	Raw material kanban	Ortiz		X			Engineered products VSIP 4.5	O'Connor		X	Shingo	■	■	■	■
3	1/15-1/19	Avon	AV-#03JanYY	Sample development standard work	Drew			X		Lead-to-land PIP action step 3.1	O'Connor		X	Shingo	■	■	■	■
4	2/5-2/9	Harwich	HA-#01FebYY	Victa production preparation (3P)	Thomas	X				Level II deliverable #17	Condon	X		n/a	■	░		
5	2/5-2/9	Harwich	HA-#01FebYY	Mistake-proof Fenn line	Quinn				X	A3 to address recurring ML320 defects/customer returns	Condon	X		n/a	░	░		
6	2/19-2/20	Chatham	CH-#01FebYY	Flow kaizen for Tornado product family	Remy	X				Need to develop future state and VSIP that compresses lead times from 24 to 12 days by 12/31/YY	Foster		X	Jeffrey	■			

Legend: ░ Started   ■ Completed

SD = strategy deployment
VSIP = value stream improvement plan
PIP = process improvement plan

Figure 4-9. Example master kaizen schedule. (For a blank form, see Appendix A.)

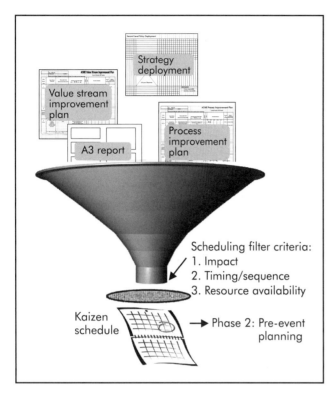

Figure 4-10. Kaizen event selection: Primary inputs and filter criteria.

and objectives are not necessarily cast in stone. The related schedule(s) has a "forecast" characteristic to it. As such, the further out in the scheduling time horizon, the less accurate it becomes.

As the lean transformation progresses, especially specific to a given value stream or critical macro process, the lean leaders and everyone else get "smarter." What was originally thought to be an excellent fourth kaizen event may now be moved to fifth or sixth, as it is determined that the third kaizen, while successful, exposed other waste that must be addressed next. For example, the more immediate need after establishing continuous flow is the replenishment and presentation of the raw materials to the assemblers within the work cell rather than the initially planned improvement on an upstream operation. Further, there may be some immediate "opportunistic" or problem-solving (for example, customer complaints) kaizen events that may take precedence. There is no dishonor in this. It just means that the schedule must be somewhat dynamic with respect to scope.

The actual placeholders and kaizen "cadence" (one major kaizen per month in a particular value stream for the first year) should not be modified. The only justifiable reasons for skipping a kaizen are in the realm of catastrophic events and tremendous abnormal demand. Canceling kaizen events usually sends the message that they are optional. Soon enough, the number of acceptable, rational reasons for postponing an event will snowball and the imperatives of strategy deployment and improvement planning will not be so imperative.

SUMMARY

- Strategy is the first phase of kaizen event standard work.
- Before conducting any kaizen, leadership must validate both the "ladder" (that kaizen is the right tool to use in this situation) and the "wall" (kaizen what matters).
- Since lean leaders typically want some quick short-term wins, in many lean launches, formal strategy deployment is delayed as much as a year or more. In these cases, the launch begins with value stream analysis and one or more value stream improvement plans.
- The three criteria for initial kaizen event selection are:

 **No filter, just schedule.** The CEO sounded just like the lean grey beard that he self-advertised, telling a senior manager to book the sensei for "a week, a month, forever." Unfortunately, that was about as strategic as it got. He and his leadership team never successfully applied the rigor of strategy deployment, value stream improvement planning, A3 thinking, or a lean management system. It was "ready, fire, aim" all the time. What was the final outcome? Liquidation.

1. impact,
2. timing/sequence, and
3. resource availability.

- Sources for high-impact kaizens include:
  - strategy deployment,
  - value stream and process improvement plans,
  - A3 reports,
  - significant gap closure,
  - opportunism,
  - the lean management system, and
  - employee suggestions.

- A long-term kaizen event schedule is a critical tool to help:
  - facilitate proper internal resource allocation/rotation,
  - secure external consultant time,
  - spread kaizen activities among different locations and functions,
  - communicate planned kaizen activities to employees, and
  - provide adequate time for pre-planning.

REFERENCE

Sobek II, Durwood K. and Smalley, Art. 2008. *Understanding A3 Thinking: A Critical Component of Toyota's PDCA Management System.* Boca Raton, FL: Productivity Press, p. 31, 62.

5
PLAN FOR SUCCESS

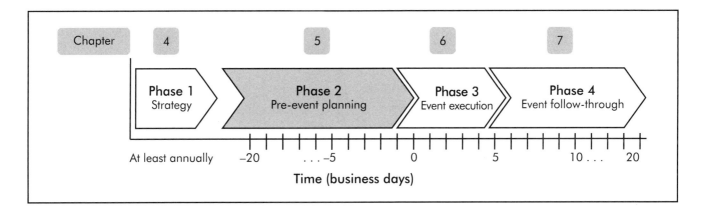

"Before anything else, preparation is the key to success."
—Alexander Graham Bell

PHASE 2: PRE-EVENT PLANNING

While kaizen espouses a "just do it now" philosophy, this is appropriate *only* in the execution phase. Pre-event planning is a purposely thoughtful process, building upon the foundation established within Phase 1: Strategy. This preparation is critical for what will ultimately be the rapid and radical improvement deployed in Phase 3: Event execution. Any oversight at the planning juncture will greatly reduce the effectiveness of the kaizen event from the perspective of achieving sustainable results and employee development and engagement.

Planning by nature is imperfect because those who conduct the planning must often work with incomplete knowledge and less than accurate assumptions. For example, value stream analysis, a mainstay of Phase 1, is by design a high-altitude tool with about 70% accuracy. However, this dynamic does not give license to shoot from the

> ### Kaizen Planning
> Should you plan to have a successful kaizen event or plan for kaizen success? Well . . . both. While this chapter is primarily about planning well to have a successful kaizen event, this can not be separated from anticipating kaizen success and planning for the implications of that success. Kaizen events can not be perceived as separate, distinct entities without linkage to anything else. Good kaizen event planning requires systems thinking. This is reflective of a profound interconnectedness within and between the various value streams. Changes in one process impacts others. This means that value stream analysis and value stream understanding are extremely important.

Lean Leaders' Short "Must Do, Cannot Fail" List
- Demand and verify that pre-event planning standard work is followed.
- Ensure pre-event planning does not design the "solution."
- Ensure that pre-work does not design the "solution."
- Review and test the kaizen event area profile for: impact, SMART aggressive objectives, linkage to strategic imperatives, proper team selection, scope, etc.

Gemba Tales

Okay, then what? I stopped in to help a team prepare for an upcoming kaizen event. I was informed that the topic was setup reduction and downtime reduction on a particular piece of equipment. When discussing the goals, it seemed that if the week were a success relative to those goals, the equipment would be up, available, and running about 75% more than the current situation. So I asked two questions: "How will the material come to the equipment that much more often?" and "How will the processed material be moved from the equipment?" Blank stares indicated that the work system had not been considered. The spot improvement to the equipment would mean little without linkage to the broader system. I asked the team to go back to the value stream map and consider the plan for success.

Tale shared by Bruce E. Thompson

hip. Instead, it should reinforce the need to follow a pre-event planning process—one that is founded upon logic and common sense and introduces a greater level of precision and specificity relative to scope, timing, objectives, team resources, logistics, etc. The kaizen event practitioner must follow the Fieldbook standard work relative to these things to realize the full potential of each event.

THE FOUR PLANNING SUB-PROCESSES

Effective pre-event planning requires the execution of four fundamental planning sub-processes as depicted in Figure 5-1. The sub-processes are constitutive of specific activities that are best conducted in a time-phased sequence. A summary level Gantt chart (see Figure 5-2) reflects the major steps, the first of which should be initiated at least three weeks before the scheduled event.

1. Event selection and definition—this represents the logical progression from and linkage to Phase 1: Strategy. It also requires a greater level of specificity relative to timing, scope, and preliminary objectives, while prompting the thoughtful selection of the team leader and team, facilitator, and/or sensei.
2. Communication—kaizen events profoundly involve and impact a number of people;

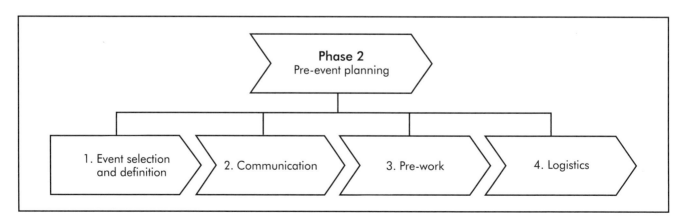

Figure 5-1. Pre-event planning: four basic sub-processes.

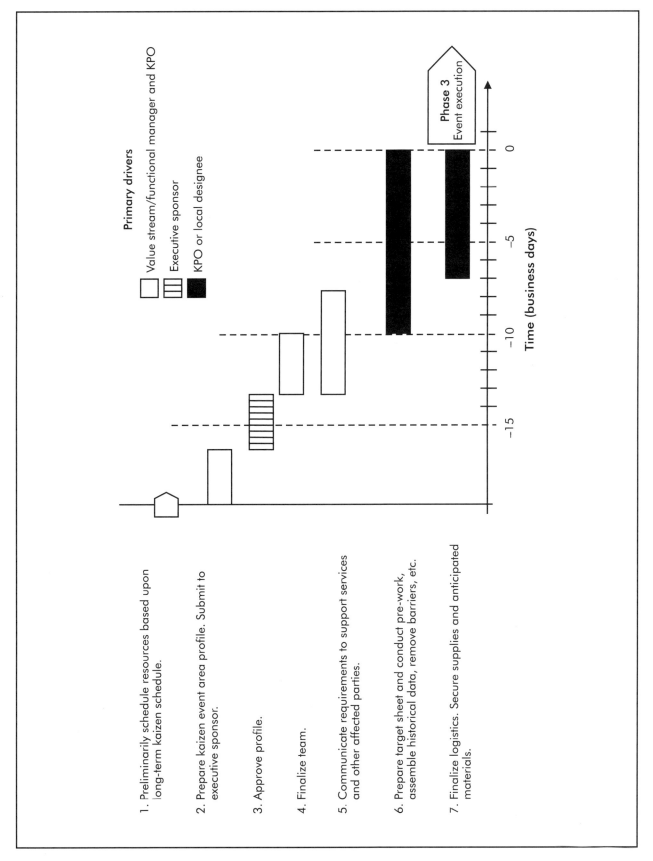

Figure 5-2. Pre-event planning sequence.

leadership, team members, suppliers, customers, visitors, upstream, downstream, targeted processes and those who work within the processes, support functions (maintenance, information technology, finance, etc.), as well as those who provide or arrange things like food, refreshments, and break-out rooms. Consistent with the notion of respect for the worker and basic change management rigor, people respond and engage best when they understand the what, who, when, how, and why of the forthcoming kaizen event.

3. Pre-work—kaizen events are about results and pre-work provides a foundation for event effectiveness. Selectively gathering data and performing analysis beforehand prepares the team members so they are not pre-occupied, de-focused, and slowed down during the event. Additionally, there may be barriers (change management, resource availability, or the like) that management must address prior to the event.

4. Logistics—the "little" things, if overlooked, can easily delay or interrupt kaizen event momentum. These things include supplies, forms, equipment, food, break-out room reservation, and petty cash.

Execution of the pre-event planning sequence and the underlying four sub-processes can be facilitated by the use of a pre-event planning checklist. The checklist, shown in Figure 5-3, should be administered by the KPO and requires the cooperation of key people such as the site leadership members, value stream manager, and kaizen event team leader(s).

EVENT SELECTION AND DEFINITION

This first pre-event planning sub-process is, in simple terms, the extension of strategy into a relatively definitive and actionable plan for the kaizen event. As reflected in the pre-event planning checklist, there are five key activities. The underlying best practices or guidance can be distilled under the following subjects:

- preliminary scheduling of resources,
- the kaizen event profile, and
- team selection.

Preliminary Scheduling of Resources

The long-term kaizen schedule(s) (see Figure 4-9) should provide a 6- to 12-month view of the calendar slots for externally and internally led kaizen activities. Phase 1: Strategy provides the real detail and heavy thinking behind what constitutes the high-impact kaizen events and their appropriate sequence. The kaizen schedule specificity should be sufficient relative to the scope, location, high-level objectives, and their linkages to things like strategy deployment and improvement planning, sponsor and KPO—the necessary inputs for the pre-event planning phase.

Senior managers should use the kaizen schedule to anticipate forthcoming events and clear their calendars for the week so that they, at a minimum, may appropriately support and attend kick-off and team leader meetings, the final report-out, and any recognition/celebration activities. Similarly, the KPO and/or designee should attend to other long lead-time items such as scheduling external consultants, which should be done as far out as 12 months in advance. A good sensei is typically in high demand and should be "locked up" on the schedule well in advance.

The long-term kaizen schedule is also useful for individuals who are looking for general kaizen participation experience regardless of scope or location. This can be applied at the personal level or extended to the training/participation of one or more of a person's direct reports. Nothing helps cultural transformation more than direct participation. It facilitates engagement, fosters support for the transformation effort, and eliminates the varied lean myths and ignorance that people sometimes harbor.

The long-term schedule can also, depending upon the specificity of the forthcoming event(s), be helpful for senior management/value stream management to anticipate possible or probable event-caused disruption of operations. Thus precautionary measures can be taken, such as scheduling overtime in and around the time of the kaizen event, modestly building inventory,

Pre-event Planning Checklist

Event dates: _____
Team 1: _____ Team 3: _____
Team 2: _____ Team 4: _____
Process owner: KPO

	Pre-event Categories		Required Activities	By Whom	Timing Recommended T-minus*	Timing Actual Completion Date	Status %
1	Event selection/ definition	1.1	Identify next kaizen opportunity and date	SLT, VSM, KPO	−30		25 50 75 100
		1.2	Verify schedule with facilitator/consultant	KPO	−45		25 50 75 100
		1.3	Identify TL, prepare profile and submit to executive sponsor	SLT, VSM, KPO, TL	−15		25 50 75 100
		1.4	Receive approved profile	VSM	−10		25 50 75 100
		1.5	Finalize team	SLT, VSM, KPO, TL	−9		25 50 75 100
2	Communication	2.1	Develop kaizen week at-a-glance schedule	KPO	−10		25 50 75 100
		2.2	Conduct event orientation meeting with team	VSM, KPO, TL	−8		25 50 75 100
		2.3	Apprise local management and out-of-town guests of event schedule, profile, etc. Ensure required management/supervisors will be present for team leader meetings and report-out	KPO	−8		25 50 75 100
		2.4	Provide specific communication to kaizen affected workers (support and targeted processes)	VSM, department manager, KPO	−7		25 50 75 100
		2.5	Provide general communication to the site	VSM	−5		25 50 75 100
3	Pre-work	3.1	Prepare target sheet	VSM, KPO, TL	−9		25 50 75 100
		3.2	Identify/address perceived barriers	VSM, SLT, KPO	−8		25 50 75 100
		3.3	Prepare/perform data collection and analysis as appropriate	KPO, TL	−5		25 50 75 100
4	Logistics	4.1	Reserve main conference room and team break-out rooms	KPO	−10		25 50 75 100
		4.2	Arrange for food and beverages	KPO	−5		25 50 75 100
		4.3	Procure/prepare recognition awards	KPO	−5		25 50 75 100
		4.4	Ensure items reflected on kaizen team supply list are present	KPO	−5		25 50 75 100
		4.5	Ensure petty cash availability, company credit card, etc.	TBD	−5		25 50 75 100
		4.6	Arrange security access/clearances for consultant and other visitors	KPO	−5		25 50 75 100

KPO = kaizen promotion officer, SLT = site leadership team, VSM = value stream manager, TL = kaizen event team leader
* = minimum number of business days from the event kick-off date

Figure 5-3. Pre-event planning checklist. (For a blank form, see Appendix A.)

> **Kaizen's "Hippocratic Oath"**
>
> Okay, the phrase "first, do no harm" was never actually in the real hippocratic oath, but if kaizen had one, that phrase should be in it. Kaizen events should never negatively impact customer service; that would be antithetical to lean.

or off-loading requirements to sister facilities or operations, etc. The lean business system, by its very nature is customer focused; missing customer requirements to conduct a kaizen event is not acceptable.

The Kaizen Event Area Profile

The kaizen event area profile (see Figure 5-4) is a cousin of the A3 report. It is a one-page document that facilitates the event selection and definition process as well as the naming of team members, facilitators, and consultants (the external sensei). The profile is the feedstock for the initial kaizen event targets and strategy and manifests the critical linkage to strategy deployment and improvement plans.

The kaizen sponsoring value stream manager, key process manager/functional head, and the supporting KPO, in conjunction with the appointed team leader, must prepare the kaizen event area profile and submit it to the executive sponsor for approval within 15 business days before the scheduled kaizen event kick-off. Clearly, there is a risk that the profile may not be approved as submitted and is instead modified. Some iterative fine-tuning should be expected, but large-scale changes are usually not a good thing.

Modification is likely to occur if the proposed event does not address priority value stream improvement items, the scope is too large or too small, it is determined that a kaizen is not the most effective/least waste way of achieving the objectives, the team composition is not adequate, etc. Because of the probability of modification, it is prudent to complete and submit the profile for review well prior to the 15-day deadline. This will provide sufficient time to regroup and refocus, if necessary.

The profile fields (Figure 5-4) should contain the following information, some of which can be directly lifted off of the master kaizen schedule (Figure 4-9):

1. *Team #*. Often a kaizen event is conducted with multiple teams. Assigning a team number, along with the event description, is an easy means to differentiate the teams in the planning, execution, follow-through, and archiving activities. Many companies will employ a coding scheme that will reflect important information such as location, team number, and date. For example, Mem-12-7/YY reflects a kaizen event in Memphis, team #12 (12th team of the year), conducted in July of 20YY.

2. *Event description*. This brief description is reflective of the event scope/area of focus and mission, for example, "Gelcoat continuous flow—mold prep to finishing." Greater detail relative to location, objectives, and scope (in scope and out of scope) are provided in other profile fields. Scope definition and selection of event objectives is part art and part science. Planners must carefully estimate the kaizen work content and related cycle time, and ensure that the cycle time does not exceed the available time to conduct the event (nominally 3 to 5 days). When properly scoped, there will be little to no open "to

> **Gemba Tales**
>
> **Kaizen in small companies.** I once ran a company with only 10 employees. Obviously, putting a team of six or seven together for a week would shut down the company, which was not a customer-focused option. However, I wanted to use kaizen to drive improvement. The solution was to create a kaizen team and allow them one hour a day for a month to work on the kaizen. We used all the same steps and tools and achieved the expected results. The only difference was we spread it over a month. We would do this every other month for a total of six kaizens a year. Being small is no excuse to not use kaizen to achieve breakthrough improvements.
>
> *Tale shared by John A. Rizzo*

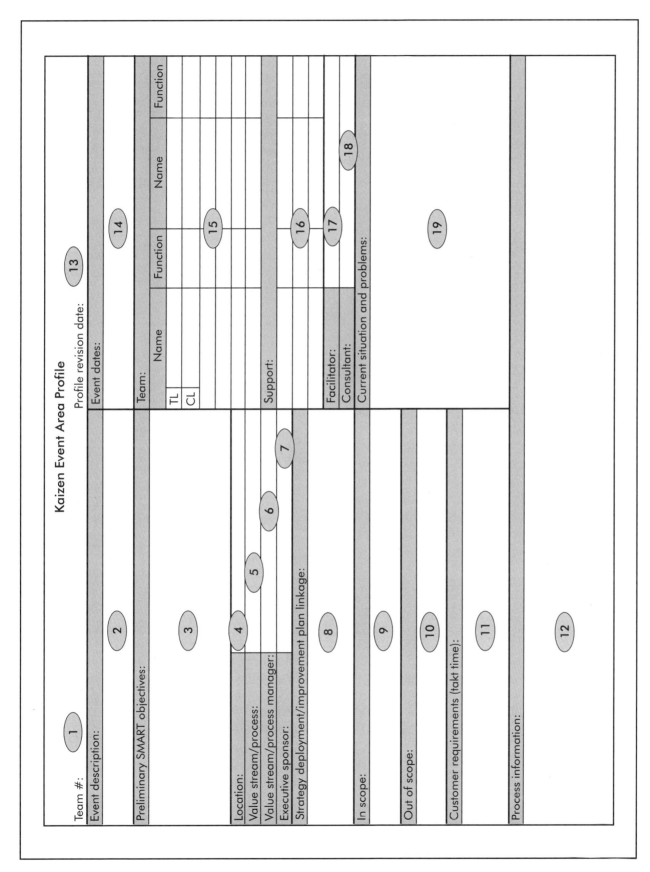

Figure 5-4. Kaizen event area profile. (For a blank form, see Appendix A.)

do" items (kaizen newspaper) at the conclusion of the event and the team will perform value-added work during the entire event. If the estimated event cycle time exceeds the available time, then the scope must be modified.

3. *Preliminary SMART objectives.* While it is likely that event objectives will be fine-tuned days later when the target sheet is prepared and during the kaizen event itself (should new and better understanding of the issues and opportunities dictate), this is where the explicit objectives of the kaizen are first stated. The preliminary objectives should be directly linked to and derived from the imperatives reflected in the company's strategy deployment, improvement plans, A3 forms, and the like, and should be SMART (specific, measurable, actionable, relevant, and time-bounded)... anything less and the kaizen may end up being little more than a study mission.

Typically, the critical few preliminary objectives should be listed in bullet point format within the profile field and use "direction/output/measure" language. For example, "Increase productivity (man hours/unit) by 30+%; establish/implement standard work; double the 5S score, etc. These preliminary objectives will be used directly to prepare the kaizen event target sheet.

> ### Specifying Objectives
> The key measures and targets should largely be "cut and pastes" from the strategy deployment deliverables, Gantt charts, improvement plans, etc. But things change, priorities sometimes shift, and management insight and understanding should get better the further the enterprise moves down the lean transformation path. Accordingly, a lean leader should not blindly accept the targets that were established sometimes as much as 6 or 9 months ago. Targets should be tested to ensure that they are still valid, relevant, challenging enough (versus too incremental), etc. It is wise to revisit the value stream maps, process maps, bowling charts, and, above all, the gemba. When in doubt, err on the side of aggressiveness!

Occasionally, such as in the design of a new process, it is beneficial for the preliminary objectives to take on a mission statement format. For example, using SMART characteristics, the objectives may read as follows, "Design a new proposal process that will be implemented by June 1; reduce RFP response time by 50%; and reduce bid generation work content by 25+% while maintaining or improving bid quality (accuracy, aesthetics, etc.)."

4. *Location.* This reflects the physical location of the kaizen event.
5. *Value stream/process.* This field captures the name of the specific value stream or key process.
6. *Value stream/process manager.* This manager is the primary owner of the value stream or key process(es) and is responsible for its day-to-day performance and improvement.
7. *Executive sponsor.* While this may sometimes be the value stream manager (possibly the plant manager), the executive sponsor is typically at least the superior of the value stream or key process manager. The sponsor should have responsibility for the strategy deployment deliverables that the kaizen will impact.
8. *Strategy deployment/improvement plan linkage.* This field should reflect the specific strategy deployment deliverable and/or the reference numbers of relevant value stream or process improvement plan objectives and related measurable goals. For example, the kaizen may be targeting the finishing loop (loop #4) with the related value stream objective, "4.1 Establish continuous flow" with measurable goal, "4.1.1 Standard WIP level of 4 units/line by Nov 15," as well as value stream objective, "4.2 Reduce work content," and measurable goal, "4.2.1 Improve productivity by 30% by Nov 15."
9. *In scope.* It is important to delineate the addressable processes, technologies, and policies that are within the kaizen team's sphere of control.
10. *Out of scope.* This presents further clarity to the team relative to scope and guards

against one of the most debilitating afflictions of any kaizen—scope creep.

11. *Customer requirements (takt time).* This field provides the rate of customer demand, typically in the language of takt time. It is important to "show the math" behind the takt time calculation to prevent confusion and more easily identify flawed assumptions (for example, the presumption that available time is not decremented for a beginning of shift "sunrise meeting").

12. *Process information.* This reflects a brief description of the process. This may be done in bullet points, a flow chart, process map, or narrative form, or it may be captured by a pre-existing and accurate standard work sheet or standard work combination sheet. Attachments are acceptable.

13. *Profile revision date.* Dutiful kaizen promotion officers have a habit of updating the profile to reflect corrections or additional information. The profile is a well-circulated document—kaizen participants, lean leaders, sensei, and key support personnel should each have a copy. The inclusion of the revision date on the profile will provide control and help avoid confusion.

14. *Event dates.* This field simply reflects the date of the actual kaizen event, for example, "12/1–12/5/YY."

15. *Team.* This provides the team roster—team member name and function. It also indicates the names of the team leader ("TL") and co-leader ("CL").

16. *Support.* A successful kaizen needs support from a wide cast of people, many of whom are not full-time team members. Typical kaizen support comes from various subject matter experts, information technology (IT), maintenance, and analysts. By specifically identifying these people at the stage of profile preparation, it is more likely that they will be forewarned, available, and engaged.

17. *Facilitator.* This reflects the name of the event facilitator.

18. *Consultant.* If the kaizen is an external one, meaning coached or facilitated by an outside

Flow Kaizen

Flow kaizen can benefit from the planning rigor implicit in the profile. Please note that product family definition will most likely require the use of a product family analysis matrix (see Glossary) to discern the product or service family(ies).

Briefly, the profile fields should be the same as those in Figure 5-4, with the following exceptions/clarification, by field reference number:

2. Reflect as "XYZ product (or service) family value stream analysis."

3. The preliminary SMART objectives are typically satisfied by:

    a. current state value stream map,
    b. future state value stream map, and
    c. value stream improvement plan.

    It is beneficial to focus the flow kaizen team by articulating that the future state value stream should, for example, reflect a certain lead time, productivity level, etc. and be accomplished by a certain date (typically 6 to 12 months from the date of the event, definitely not exceeding 18 months).

8. While the VSIP linkage is obviously not applicable, the strategy deployment linkage may be very relevant.

9. This can easily define the boundaries of the value stream analysis (that is, order to delivery, concept to new product launch, etc.)

11. Calculate product family's relevant takt time.

12. Not applicable.

13. The nominal flow kaizen duration is 4 to 5 days.

15. The team should be comprised primarily of the management team responsible for the product or service family value stream.

16. Flow kaizens are essentially "paper kaizens." Therefore, support members will be predominately subject matter experts.

19. Briefly describe the current situation and problems; because this is at the product or service family level, the measurements and descriptions should be at an appropriate macro level.

consultant, this profile field should provide the name of the sensei. If it is an internal one facilitated by an internal resource(s), the field should reflect "N/A."

19. *Current situation and problems.* This portion of the profile should provide a brief description of the current condition that is relevant to the kaizen and its scope. For example, "Setup time exceeds 3.5 hours, resulting in an 'every part every interval' of once/two weeks and significant batching and queuing with 1.5+ months of WIP . . ." The problem description may be reflected in bullet point or narrative form and supplemented with standard work sheets, standard work combination sheets, % load charts, histograms, run charts, etc. Attachments are acceptable and encouraged assuming that the data is relevant and accurate.

Team Selection

The selection of the team must be done before submitting the profile for review. However, with the exception of the identification of the team leader, with whom the profile should be completed, team membership should be the last entry on the profile. In other words, the initial pre-event planning focus must be to identify the high-impact event "candidate" and related issues, and then the focus shifts to determining the team that will best support a successful event. Build the team around the event, not the event around the team.

As mentioned, the team leader has a role to play in the event selection and definition process. He or she should directly participate in the preparation of the profile. This will foster an explicit understanding of management expectations and a profound level of engagement and ownership. As part of this front-end process, the team leader should exert strong influence in the selection of team members. However, before exploring team member selection, there must be an explanation of the criteria for team leader and co-leader selection.

Team Leader and Co-leader

Each kaizen team should have one team leader and one co-leader. The co-leader should either possess or have real and near-term potential for developing team leader skills. While for simplicity this section focuses on the team leader role, it is important to note one of the most important distinctions between the two roles relates to their regular, non-kaizen roles and responsibilities.

The team leader should not have day-to-day responsibilities within the kaizen target area. This "neutrality" is important so that bias is not introduced during the course of the event. In contrast, the co-leader should, if possible, be from the target area. The co-leader, by virtue of his or her explicit stakeholder position, should facilitate the training and transition to the newly implemented standard work during the event. Afterward, he or she then helps to ensure that the gains are sustained.

The selection of the event team leader is extremely critical, particularly in the early stages of an enterprise's lean transformation when most kaizen participants have between little and zero kaizen experience. In short, a sub-par team leader can seriously marginalize an otherwise well-planned event. This is not a decision where caution is thrown to the wind.

Leadership skills are quickly tested in the kaizen event environment where, by design, the nature of the decision cycles (plan-do-check-act [PDCA]) is accelerated, compressed, and repeated many times over in a span of 3 to 5 days. This can be exacerbated by a number of other dynamics, including:

- the pressure to implement improvements;
- senior leadership "attention";
- the application of philosophy, tools, and techniques that may be foreign to many; and
- the likelihood that the output will effect a change in the way people work and are measured.

As if this "whirlwind" is not enough, the balance of the kaizen team is also experiencing much the same and responding in a variety of different ways based upon their temperament, experience level, and degree of engagement.

Leadership weaknesses may be compensated by the inclusion of strong team members. However, team dynamics often will evolve such that the stronger members will become the "de facto" leader(s). This obviously is not the most effective strategy. What are the considerations when selecting a team leader and co-leader? Figure 5-5 focuses on three areas:

1. core competencies,
2. technical competencies, and
3. the candidate's organizational role and responsibility (relative to the kaizen scope).

Core competencies. Core competencies represent a suite of "basic utility" performance skills used in a broad range of jobs. They are largely comprised of work habits, attitudes, behaviors, and personal characteristics, which dictate "how" a person performs his or her job. This is in contrast to the "what" or job content reflected within the regime of technical competencies. The kaizen event team leader and co-leader must possess both types of competencies. But while technical competencies, consistent with the lean learning model, are predominately learned by doing, core competencies are much tougher to develop or shift in the short term.

In the quick, rough and tumble world of kaizen, if the team leader or co-leader is struggling with the basic "how" of leading, he or she will most likely be behind the curve for the entire execution phase and the team's performance will suffer. Figure 5-6 highlights the relevant leader core competencies and gives examples of the situational requirements.

In many ways the team must "gel" in a matter of hours if it is to successfully negotiate the challenges presented between the kick-off on day one and the report-out on day five (or sooner). The campaign orientation of the kaizen, along with its focus, pressure, shared break-out room, action orientation, real-time feedback, celebration/recognition of each win, and redoubling of effort after each setback provides an opportunity to develop a deep camaraderie and sense of purpose. The four phases of team development: forming, storming, "norming," and performing, must be negotiated at hyper-speed.

Technical competencies. The team leader does not have to possess a Ph.D. in kaizen. The technical skills will develop with direct

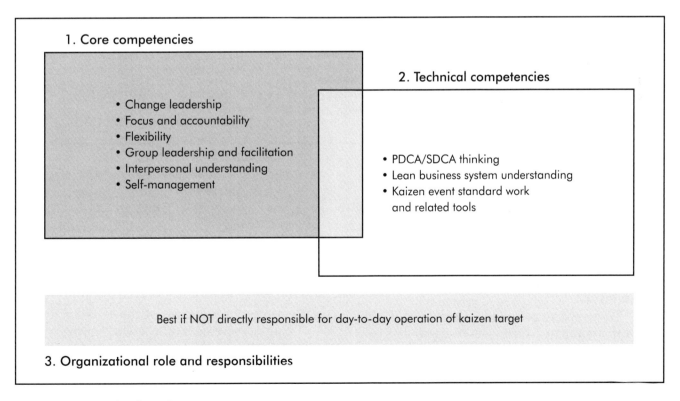

Figure 5-5. Team leader selection criteria.

Competency	Description
Change leadership	The leader must believe the kaizen process is an effective and substantial part of the enterprise's business strategy. Through all phases of the kaizen event, the leader must ensure that team members, kaizen affected employees, and others understand the reason for the kaizen event, the approach, impact, as well as new expectations relative to accountability and behavior.
Focus and accountability	The leader must zealously follow the rigor of the kaizen event standard work and internalize the mantra of: 1) identify the waste, 2) acknowledge the waste, and 3) eliminate the waste. During the event, the leader must move the team to implement as many improvements as possible. A proper mix of collaboration, salesmanship, and blunt orders is needed to get the job done. The leader must have the staying power to effectively work long hours while leading and exhorting others to do the same. This same stamina must carry through to the post-kaizen follow-up.
Flexibility	The kaizen event is a microcosm of the continuous improvement PDCA cycle. Feedback is swift as is the decision process and course correction. The leader needs to be highly flexible to excel in such an environment. He or she also must be able to accept constructive criticism and guidance from team members, management, facilitators, and sensei.
Self-management	The leader must serve as coach and facilitator to effectively engage the team members and harness their energy and creative powers. Non-participative styles dampen team spirit, buy-in, and results. The team leader must model effective team member behaviors and practices.
Group leadership and facilitation / Interpersonal understanding	The leader must provide the team members with direction and focus if they are to effectively attack the objectives and stay within the event scope. Time is short in the event; all team members must be leveraged to achieve the objectives and accomplish as much as possible. This requires the ability to strategize, "read" individual work styles, and then follow with separation and delegation of tasks among the team members. Time pressures and those brought about by challenging the status quo often require conflict resolution skills.

Figure 5-6. Team leader core competencies.

experience, study, and reflection. That said, there is an expectation that team leaders who work in a company that has moved well down the lean transformation path will have had real previous multi-kaizen experience as at least a participant and co-leader.

For those who are called to serve as a team leader in a company that is just beginning its lean launch, there are a few prerequisites. These prerequisites include a well-developed intellectual curiosity, the gumption to do some pre-kaizen studying, and the willingness and humility to learn the rest during the kaizen from the sensei and others. The new team leader should, at a minimum, possess:

- a basic understanding of the scientific method (PDCA/SDCA);
- a general awareness of the construct, principles, systems, and tools of the lean business system model; and
- a familiarity with the standard work underlying the kaizen event—its steps, sequence, and the tools as manifested in the most

> **No Yo-yos**
>
> Make no mistake. Team member commitment must be full time for the event. Exceptions to this rule should be extremely rare. "Yo-yo" team members are those who are repeatedly pulled out of the event for meetings and projects by their superiors. These members accomplish little, rob the team of resources, disrupt team progress, and demoralize the truly dedicated team members. In short, Yo-yos take a spot on the roster that would have been better filled by a more permanent team member. Effective lean leaders do not "pull the string."

basic forms and techniques—time observation forms, standard work sheets, standard work combination sheets, percent load charts, kaizen target sheets, kaizen newspaper forms, improvement idea forms, and process mapping techniques (see Chapter 6).

Multi-team kaizen events typically mean that sensei and facilitator availability is diminished as they are spread across more teams and more "students." In such a situation, it is more important to have enlisted team leaders who have greater technical depth, not to mention stronger group leadership and facilitation skills. To mitigate risk, it also makes sense to conduct more comprehensive pre-work and prepare a thoughtful initial kaizen strategy.

Organizational role and responsibility. To avoid or minimize bias, typically the team leader should be from an area or process outside of the kaizen scope. Co-leaders, however, should be from the area or process within the scope of the kaizen as they will provide detailed insight into the related processes and people.

After an area has conducted a few events, experienced team leaders should act as co-leaders to help develop and test new team leaders. The capable co-leader is there to pick up the fumbles and keep the facilitator (or consultant) informed. The facilitator, in turn, will coach the new leader on the proper styles and techniques as necessary.

Better Selection = Better Results

The team leader, in conjunction with management, should select the team. This selection should be influenced first and foremost by the mission at hand, which is crystallized in the profile.

To best fit the team to the particular and relatively short-term team mission while remaining mindful of the more macro and longer-term mission of the enterprise's lean transformation, every team member must be chosen for a purpose. This purpose is multi-dimensional and extends to things like the candidate's core and technical competencies, organizational position (both formal and informal), and needed exposure and immersion into lean and team skills. The most basic considerations are team:

- size,
- composition and chemistry, and
- specific roles and responsibilities.

A secondary, but extremely important consideration, is team training.

Team size. The ideal kaizen team size is roughly six to eight members (Figure 5-7). This size accommodates proper team composition from the perspective of required skill sets, kaizen experience, roles and responsibilities, change management opportunity, and cross-functional representation. It also facilitates good group dynamics, including the all-important problem-solving and, ultimately, *results*.

Teams with fewer than six members often require a reduction in kaizen scope to ensure that the work can be completed by the end of the event. A typical problem with small teams is manifested toward the end of the event, when there is no remaining time to validate "improvements" and document the new standard work.

> **One Perspective**
>
> Jeff Bezos may know little about kaizen, but he is not too far from the conventional wisdom.
>
> "If you can't feed a team with two pizzas, the size of the team is too large."
>
> —Jeff Bezos, Chairman, CEO, and Founder of Amazon.com

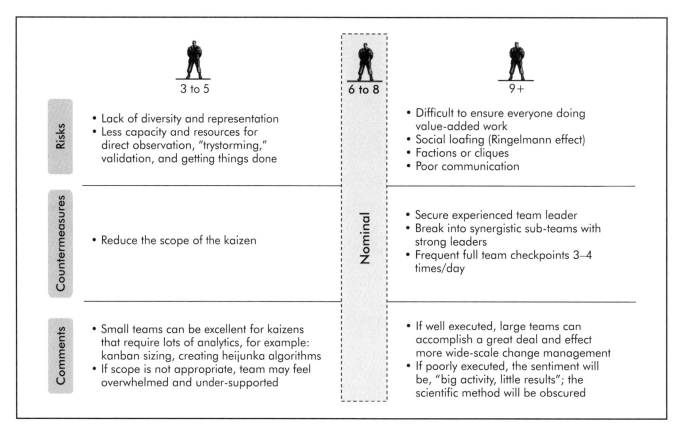

Figure 5-7. Team size.

This is clearly a recipe for post-kaizen sustainability problems.

Teams with more than eight members require a seasoned kaizen team leader who is knowledgeable in the application of lean principles and proficient at team management. Large teams increase the probability of two types of team muda: 1) "social loafing" and 2) lack of direction. The former is reflective of the "Ringelmann effect" and the inclination of participants to slack and hide because they can. The latter is where team members, who, by no fault of their own, experience firsthand the wastes of waiting, over-processing, or event defects. Of course, over-processing also increases the risk of "scope creep" where underutilized team members foray into untargeted parts of the value stream or beyond.

With a large team, it is often difficult to ensure that all members are performing value-added work. Good team members genuinely desire to move the ball forward and can easily discern "busy work" from value-added work. They understandably get frustrated when they must wait for needed direction or are allowed to proceed far down a path without a necessary checkpoint from the team leader or facilitator, only to find out that the work was a duplication of what another sub-team had already done or a large portion is wrong and must be redone or reworked. It is no surprise that as team members experience this dynamic, and/or they are repeatedly saddled with mundane tasks, they are more inclined to practice some social loafing.

A properly planned and executed event will be a positive experience for each team member and reinforce the value of lean thinking. Conversely, team members who have a negative experience will disengage and, no doubt, share their thoughts with co-workers.

Team composition and chemistry. As depicted in Figure 5-8, the team composition and resultant chemistry are determined by four basic criteria:

> **Social Loafing**
>
> Maximilian Ringelmann, a 19th century French agricultural engineer, discovered that as more people pulled on a rope, not surprisingly, more force was imparted. However, the increase in the force was not commensurate. Indeed, he measured a type of "social loafing"—the individual, per capita, effort lessened as people were added!

1. representation,
2. technical competencies,
3. core competencies, and
4. kaizen experience.

None are truly mutually exclusive, especially in the context of a team of six to eight individuals.

Balanced team representation promotes diversity, perspective, and ownership. There are three generic areas from which to choose:

1. the target area itself,
2. upstream and downstream processes from the target area, and
3. "fresh eyes."

Each constituency should be roughly equally represented (see Figure 5-9).

Team members from the target area include management/supervision and workers (designated as the "operator" role). Their balanced participation typically addresses at least two of the selection criteria, representation and technical competencies, specifically process expertise. Their inclusion also addresses change management opportunities. A profound level of engagement in the kaizen and the resultant standard work usually translates into ownership and sustainability.

Upstream and downstream processes represent either those processes directly "adjacent," meaning they directly supply the target area or are a direct customer, or are still further upstream or

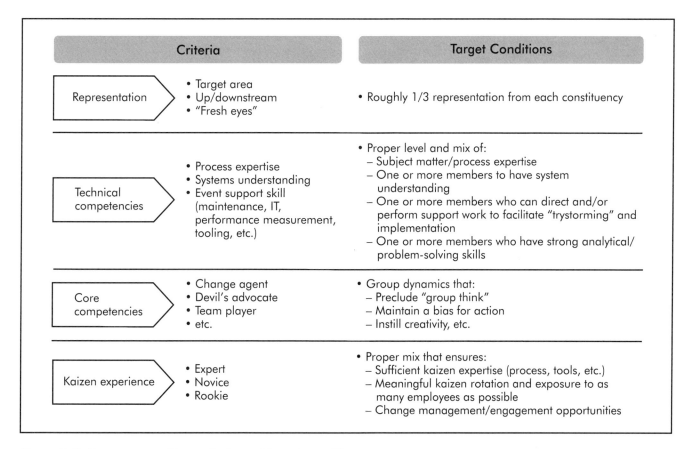

Figure 5-8. Team composition: criteria and target conditions.

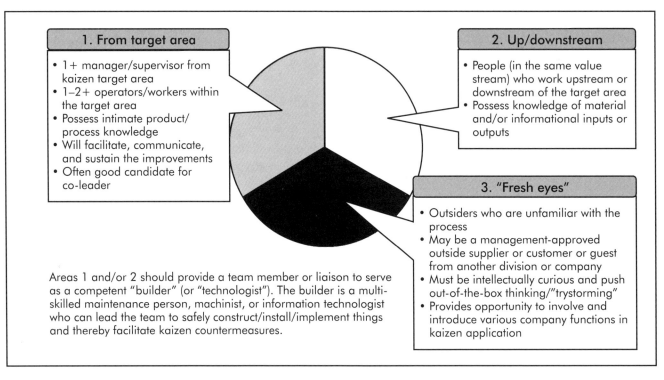

Figure 5-9. Team representation: rule of thumb.

No kaizen without operator representation. Being new to an organization that had a solid lean pedigree, I was shocked to go to a kaizen report-out and find that the event had been held the night before, after the work shift, and that *no* associates from the work cell were involved. In fact, no shop-floor associates were in the event at all. During this late afternoon "stealth" event, the work cell was moved from one part of the shop floor to another, the cell layout was changed entirely, and associates' personal belongings were moved and put into new bins. All of this was done with no input from the people who worked in the cell. They were not even aware that an event was scheduled. Needless to say, the cell associates were furious and would have no part of "standard work" after that. As the VP of engineering at that time, I swore that no one within my department would ever participate in an event unless the workers were sufficiently represented on the kaizen team.
Tale shared by Evan M. Berns

downstream of the target. Members from the up/downstream constituency will provide valuable insight into inputs and outputs and, with substantive knowledge of the target process, may have an objective perspective on things like process performance, requirements and information flows, and linkages to other processes. Depending upon the nature of the event and the lean maturity of the company, the more remote processes can extend to the external suppliers of goods or services and customers at the end of the value stream. External suppliers and customers can be categorized under "fresh eyes."

The third and last, but definitely not least, constituency is comprised of people with "fresh eyes." These are people who are not normally associated with the kaizen target area and many times are lean neophytes. They get to ask the common sense ("dumb") questions like, "Why do you do it that way?" because, unlike many other people, they are not encumbered by habit, direct ownership, and political and cultural stakes. Their typically fresh and unbiased perspective is further enhanced if it is accompanied by a deep intellectual curiosity and an outspoken nature. "Fresh eyes" often hail from functions such as program management, human resources, and finance.

External suppliers and customers can serve two roles. They bring a much different perspective and can often provide unique help in problem-solving or countermeasure activities. External customer participation should have prior senior management approval. For obvious reasons, before such an invitation is extended, there are some prerequisites or conditions that should be satisfied. For example:

- the host company will want to have achieved some demonstrated level of proficiency in kaizen (poorly executed kaizen events in the presence of the customer are not very impressive),
- team members must understand how to tactfully engage with the customer, and
- the customer must be more inclined to partnering than extorting one-sided gains.

Kaizen, conducted properly, will expose problems in the most objective and sometimes brutal way. A company must be prudent as to when, where, and how customers participate in in-house kaizens, especially when the company's lean transformation status is immature.

Technical competencies. A strong kaizen team possesses all three of the following technical competencies:

1. process expertise,
2. systems understanding, and
3. event support.

These are by no means mutually exclusive.

The more complex the target process, the more process expertise is required on the kaizen team. While direct observation is a foundational element of all kaizen, it does not eliminate the need to have deep process understanding. A team should not attempt to "improve" what they do not understand.

Process expertise should come from multiple levels and sources, including the workers or operators who do the job on a daily basis, as well as those who provide technical support.

Gemba Tales

"Fresh eyes" and 5 whys. The kaizen team's mission was to design a process for a stainless steel part that would reduce its lead time from 5 weeks to "make today everything the customer ordered yesterday." After consolidating the processes previously shared among five functional departments down into a cell with single-piece flow, the lead time was reduced to 6 days. Five of these days were for outside finishing.

One of the team members was a division accountant. He knew nothing about the product and little about manufacturing. He asked, "Why is a stainless steel product being treated?" He was told to prevent rust. He persisted, "Why would we worry about rust?" He was told a few customers required the finish. He asked, "Why would the customer care?" He was told they had rust on the product. He asked, "How could they have rust on a stainless steel product?" After a day of investigation it turned out the rust was from the residual metal that was left on the product from machining. He asked, "How can we eliminate the residual rust caused by the tooling?" With a higher-grade cutting tool and a right-sized washer in the cell the team was able to eliminate the outside treatment. This saved $60,000 a year in outside finishing costs and reduced the lead time to one day. If not for the "fresh eyes" asking the 5 whys, the team would not have hit a home run.

Tale shared by John A. Rizzo

Technical support includes people who provide sustaining engineering to the process, product, service, or system infrastructure, and those who provide after-market service. These roles often have generic titles of engineer, technologist, or manager preceded by "process," "product," "service," "program," or "project" in one of the following disciplines: manufacturing, software, mechanical, electrical, chemical, industrial, design, etc. The important thing is not the title, but the (shareable) knowledge. Many times, the person who does not possess the formal and fancy title may have the most developed, time-tested process knowledge. Assuming that there is no fatal deficiency in behavioral skills, this is the person(s) who should be on the kaizen team. One added benefit of having technical people on the team is that they are often, by trade, excellent problem solvers.

Consistent with the concept of a flow kaizen driving process kaizens, it is important to have one or more people on the team who have a solid understanding of the underlying system and its linkage to upstream and downstream processes. Without such a perspective and sensitivity, there is a risk that the kaizen team will optimize the target process while sub-optimizing the value stream.

Event support skills are often overlooked until the kaizen is already in process and there is the late recognition that a certain expertise is required to make one or more key countermeasures a reality. "Out of team" experiences are painful, especially when they result in the abrupt stop or redirection of the team's momentum as the countermeasures are handed off to someone outside of the team to complete. These outsiders do not always share the same sense of urgency, insight, or buy-in as the team members—often resulting in long lead times and poor quality. For these reasons, it is important to anticipate the required event support skills and ensure that one or more members possess those skills or at least can function as an effective liaison(s). These individuals can communicate with and direct and influence those who are not part of the team. Sometimes these team members are called "builders" or "technologists."

Core competencies. Core competencies are the primary drivers of team chemistry and group dynamics. Team member candidates must be assessed as a whole relative to representation, technical and core competencies, and kaizen experience. The biggest deal breaker is often a deficiency in core competencies, sometimes (incompletely) referred to as "behavioral skills." Virtually everything else, more or less, can be compensated for.

Similarly, the team must be selected so that the combination of the various member core competencies and temperaments is favorable. Contrary to some conventional wisdom, there is a place for good-willed conflict and contention. This dynamic will help ensure that the team challenges the status quo and pushes for radical improvement.

Team members should share some of the same core competencies as reflected in Figure 5-5. One or two members, but no more, should fit into a beneficent devil's advocate role. Creativity is also an important needed skill, but those chosen should not be of the "mad scientist" type that has no respect for the PDCA rigor of kaizen.

Kaizen experience. The profile explicitly reflects the short-term kaizen mission, while the broader lean implementation journey is long-term in nature. There is a definite intersection of the short-term and long-term in each and every kaizen and, with that, each team selection activity. One important question, among many, is, "How can kaizen team selection help drive cultural transformation?"

The workforce is the engine of the lean business system. Assuming that leadership does its job, a major seedbed for workforce engagement is broad kaizen participation, which:

- facilitates buy-in to the kaizen process and philosophy;
- educates people on the scientific approach of kaizen and its techniques;
- "unfreezes" workers by encouraging and enabling them to make changes;
- facilitates team work;
- helps to better communicate the "gospel"; and
- ensures faster, more far-reaching and sustainable improvements.

Beware!

Five conditions that foster "group think":

1. Group dynamics are characterized by amiability and esprit de corps.
2. The team (or de facto) leader is powerful, opinionated, and vocal.
3. Team members operate under considerable stress.
4. Team members have a strong desire to conform to certain cultural norms.
5. There is no explicit decision-making process.

(Stuart 2007)

The objective should be to rotate employees so that there may be as much as 2/3 or more of the workforce participating each year in kaizen events.

Leadership should strategically involve influential leaders (formal and informal) who are yet undecided about lean but, if convinced, would be substantial transformation allies. These people may be part of a team assigned to a "target rich" environment—one that has a high potential for process improvement.

While rotation is important, it must be done prudently. Nothing breeds success, like success. Except at the beginning of a lean launch (the first kaizen), teams should always have several members with kaizen experience. This experience includes an aptitude with the forms and techniques as well as a firm grasp of kaizen expectations, a requisite sense of urgency, and a bias for action.

Roles and responsibilities. Teams are most effective when roles and responsibilities are clearly identified, communicated, and internalized. This is even more critical in the dynamic environment of a kaizen event. While the immediate focus is on event planning and execution, the practitioner, especially leadership, should understand the long-term benefits that accrue from role clarity and discipline—the development and grooming of future leaders and team members.

The traditional kaizen team roster should reflect certain roles, some of which may be combined depending upon team size, target scope, and member skill sets. The following is a list that *characterizes* kaizen event team member roles and responsibilities (see Table 5-1 for insight):

- team leader,
- co-leader,
- navigator,
- "fresh eyes,"
- operator,
- builder/technologist, and
- compliance officer.

The roles and responsibilities of the team leader and co-leader were introduced previously as were the notions of "operator," "fresh eyes," and "builder/technologist." The role of "navigator," also (unfortunately) called "scribe," serves an extremely important purpose—ensuring the accurate, complete, sequential, and timely preparation

> **Kaizen Participation Transforms the Culture**
>
> Thedacare, a Wisconsin-based healthcare group, has correlated event participation with lean buy-in. The group's data reflects that it takes two weeks of kaizen event experience before there is any positive, meaningful register on member survey scores. This favorable trend continues up through the eighth week of kaizen participation, upon which, ostensibly, the experienced kaizen team member has achieved a significant level of self-transformation and lean understanding. Clearly, it will take a number of years before everyone in the workforce can approach this level of kaizen experience, but it is a critical element of lean transformation success and one that should be actively pursued (Koenigsaecker 2007).

Gemba Tales

From disgruntled employee to change agent. It was mid-kaizen by the time the senior operations manager was able to come to the site to observe and provide support. By the last day of the event, she pulled the sensei to the side and expressed her amazement regarding Bill's transformation. Prior to the kaizen, Bill had been very disgruntled. This had manifested itself in a number of ways, including frequent e-mails to senior management complaining about how flawed the business and underlying processes were.

The kaizen experience was a sort of metamorphosis for Bill. After a day or so of tentative kaizen participation, he saw the opportunity to significantly improve the litigation process and fully engaged. When the sensei challenged the team for a visual control to support the new process, it was met by more than a few not atypical reasons why it could not work. Bill, however, conceptualized an approach, mocked it up, and sold the team on its benefits. By the last day of the event, he vigorously and compellingly insisted that he and his staff be one of two pilot teams to prove out the new process.

Table 5-1. Kaizen team roles and responsibilities

Role	Prerequisites	Pre-event	Kaizen Duties During Event	Post-event
	All must have positive attitude, open eyes to the process and open mind to each other	All must prepare to spend full time on kaizen	All full time on kaizen—share work, documentation, and cleanup	All must clean up
Leader	• Clearest vision of objectives and how they affect the company • Understands systems thinking • Team-building skills • Expert in kaizen	• Participates in pre-planning • Researches cost, quality, delivery issues • Prepares target sheet	• Lead team to gemba • Study current situation using key measures • Try out improvements • Compare improvements using key measures • Try again and again and again • Keep things moving, ensure every member contributes to best of his ability and knowledge	• Lead final report-out • Lead role in follow-through phase • Make it stick
Co-leader	• Deep understanding of safety • Understands systems thinking • Team-building skills • Expert in kaizen	Researches area safety issues	• Ensure compliance with all safety rules during kaizen • Refer to profile sheet every day • Ensure lunch is ordered	• Lead posting of visual controls • Post final profile and target sheet
Navigator	• Competent to document standard work • Competent with kaizen format and forms • Able to capture information and drawings • Expert in kaizen	Verifies supply inventory	• Leader of standard work • Check daily progress vs. target sheet • Record countermeasures • Lead team to document before and after using standard operations forms • Lead team to make standard work visible	• Post-kaizen newspaper • Get team picture • Archive event results and documentation
"Fresh eyes"			• Ask why at least 5 times • Help team find "out-of-the-box" ideas	• Share the experience with others • Evangelize those back "home"
Operator	Firsthand knowledge of kaizen target area	Complete training as needed	Communicate and try new ideas	• Share experience with others • Help sustain gains
Builder/ technologist	Multi-skilled maintenance person, machinist, IT support person, etc.	Set aside anticipated supplies and tools for kaizen	• Lead team to safely make, modify, move, and test things for the kaizen • Serve as liaison with support functions (electrical, plumbing, IT, finance, etc.) and outside contractors	Ensure audit sheets are in place
Compliance officer	• Product/service and process knowledgeable • Knows and complies with regulatory requirements	Set aside compliance orientation material and extra personal protective equipment	• Conduct safety orientation for all visitors • Ensure all changes are compliant	Audit regulatory compliance to standard work, work area design, etc.

of the standard operations and other kaizen forms (for example, documenting improvement ideas, the kaizen newspaper, etc.). This is much more than a secretarial function; it facilitates adherence to the underlying kaizen standard work while ensuring accountability and follow-through.

All team members are expected to conduct themselves and their activities in accordance with the relevant environmental, health, and safety (EH&S) code of conduct rules and regulations and change management policies. Self-management is expected; however, it is prudent to assign a specific compliance officer role to one (or more, depending upon the breadth) team member. Kaizen events, by definition, drive rapid change, but everything must be done safely during the event and this extends to the post-kaizen condition. Safety and regulatory compliance are non-negotiable!

The compliance officer should ensure that all team members are appropriately oriented/trained and compliant in EH&S matters (for example, personal protective equipment, relevant emergency action plans, etc.) during the event. This is especially important as the team brainstorms, "trystorms," and ultimately implements its countermeasures. The resultant fabrication, demolition, movement, startup, etc., whether conducted by team members, other company personnel, visitors, or outside contractors, must be done safely and efficiently. Care must be taken to follow the company's management of change process, including proper decontamination, proper disposal of hazardous materials, use of hot work permits, and proper lockout/tagout procedures. Of course, any new standard work, by definition, must be safe and ergonomically favorable. The compliance officer's scope should also extend to areas covered by legislation such as the Sarbanes-Oxley Act of 2002.

Team training. For team members to perform to the best of their ability, they must be properly trained. Each member should receive a substantive introductory overview of lean on the order of at least several hours. This training is best conducted in the week prior to the event, but is often part of a site-wide lean launch orientation.

Additional, tool-specific training (for example, on standard work, setup reduction, production preparation process [3P], process mapping, value stream analysis, etc.) tailored to the kaizen itself also can be delivered in the week prior, but is typically done on the first day of the kaizen event (see Chapter 6).

A short overview of the kaizen process/standard work, roles and responsibilities, and team ground rules usually accompanies tool-specific training. It is helpful for team members to have their own personal, summary level reference materials in corner-stapled or plastic spiral bound format. Such materials provide meaningful examples and direction as team members try to employ the newly learned lean concepts. Throughout the event, the sensei or facilitator will provide formal or informal just-in-time training as new needs and opportunities are identified.

Roles of Facilitator and Consultant

Clarity in team member roles and responsibilities is critical to the success of the kaizen and development of team members and, ultimately, the organization. The same is true relative to the facilitator and the consultant, especially in the team's understanding of these enabling and coaching roles.

While there is an appreciable difference between the roles of a facilitator and consultant, (see Figure 5-10), it is useful to first make a distinction between them and the kaizen team. The facilitator and consultant, by and large, drive and teach the kaizen team members so that they can be more effective relative to their kaizen targets specifically (short term) and lean thinking in general (long term). Facilitators and consultants will "roll up their sleeves" and work with the teams, but they are not the doers. Only the team, as enabled and empowered by company leadership, can make it happen. This is not a dependency model. The ultimate objective, like in any lean application, is for the team to effectively self-manage as much as possible within an environment of focus, learning, rigor, and sense of urgency.

Facilitator. The facilitator, typically the kaizen promotion officer from the on-site lean or kaizen promotion office (KPO) function, is the internal resource who coordinates the kaizen

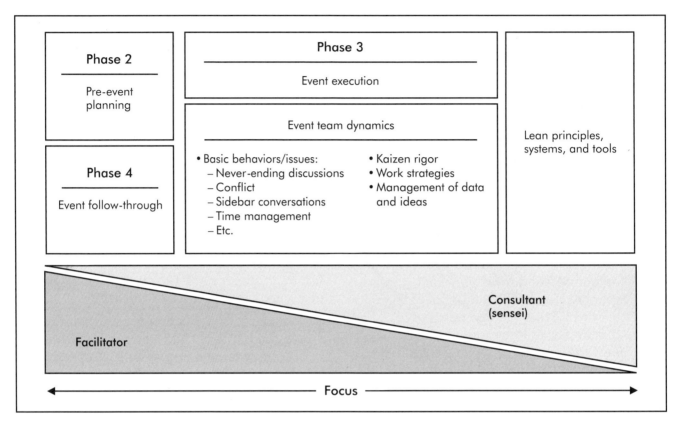

Figure 5-10. Roles of facilitator and consultant.

process. This process encompasses phases two though four of the kaizen event multi-phase approach: pre-event planning, event execution, and event follow-through. The facilitator ensures, for example:

- the items reflected in the pre-event planning checklist are completed in a timely and appropriate manner;
- the activities listed within the kaizen week schedule at-a-glance are occurring as planned;
- management is providing the necessary support and is in attendance at the appropriate meetings (kick-off, team leader meetings, and report-out);
- required supplies are on hand and requested ones (within an established dollar limit) are funded (order and retrieval is often left to the individual teams or other support staff such as purchasing or the location's designated "runner");
- final presentation materials are handed in by the team leaders and archived; and
- kaizen newspaper forms are posted, reviewed, and completed (in accordance with the local leader's standard work).

A facilitator also may be needed, depending upon the computer capabilities of the teams/team members, to generate simple signs and visual controls. The conceptualization, lamination, and installation of the visual controls are up to the teams themselves.

The facilitator is also often a sensei in training, learning from an outside or internal consultant (see Chapters 6 and 8). When no consultant is assigned, the facilitator also serves as the sensei. In this situation, it makes sense to assign a kaizen coordinator to handle the event logistics. This will ensure that the facilitator can focus on guiding the team(s) through the scientific process of the kaizen event itself, rather than being pre-occupied with things like lunch arrangements and dwindling supplies of flip charts.

Consultant. During the event, the consultant serves as the sensei, teacher, coach, or mentor for the kaizen teams (anywhere from one to four teams can be handled comfortably by the consultant, based upon his experience) most immediately but also quite necessarily for the leadership. The sensei must be an expert in lean principles, systems, and tools, as well as the kaizen process. This experience should be developed through practical application as a successful senior leader in the lean transformation of one or more operations and the coaching/facilitation of at least 100 kaizen events.

As one man was wont to say using the language reminiscent of biblical prophets, "The consultant comes to comfort the afflicted and to afflict the comfortable." The consultant, based upon experience, knows that the teams and the organization can go well beyond where they presently are and think they can go. Accordingly, the consultant will push, challenge, cajole, and furnish solicited and unsolicited feedback to the point of bluntness to move the team(s) forward. This forward movement is for the purpose of achieving the kaizen targets as well as training and development. Because the kaizen embodies the scientific process that is so inherent to lean, the consultant will teach, then require and, if need be, demand that the event standard work is adhered to.

Role of Management

Transformation leadership (Chapter 3) is essential to a successful lean transformation. The same extends to and under-girds the kaizen event. However, if management's role in each kaizen event were distilled into the most basic requirements, it would yield the following straightforward, but not necessarily "easy," things. Management should ensure that:

- kaizen events are properly selected, yielding high-impact results intimately linked to the company's strategic imperatives and value stream transformation objectives;
- the standard work reflected within the Fieldbook is adhered to;
- proper and effective communication is made to the workforce stressing company commitment, proof of the need, etc.;
- necessary resources are provided to the kaizen team to facilitate success;
- expectations from participants in regard to achievement, philosophy, and level of effort are communicated explicitly;
- a blame-free environment exists to promote positive, sustainable change;
- full-time workers are assured that employment will not be lost as a result of productivity improvements;
- there is a clear plan for the redeployment of liberated workers and their value-added assignments; and
- expectations relative to the non-negotiable workforce adoption of verified standard work/processes (until it is improved again and captured in the next-generation standard work) are explicit.

Management must understand that nothing communicates commitment, support, and a sense of urgency more than personally participating in events, being present at the kick-off meetings and the final report-out, and actively engaging in the team leader meetings. These responsibilities must be considered leader standard work for each

---

Facilitation Style—Scary or Human?

There are two basic kaizen facilitation styles. One, identified by some as "suzumura style" and ostensibly named after a disciple of Taiichi Ohno, Kikuo Suzumura, means "scary style." This style is largely defined by the following characteristics: "strict, demanding, short-tempered, insulting, and demeaning." In contrast, the other style, "Cho-san style," meaning "human style" is named after Fujio Cho, Chairman of Toyota Motor Company and practitioner of kaizen. While still demanding, it incorporates an even temper, respect, humility, benevolence, and humor. Both styles are oriented to achieve the same ultimate objectives. The effectiveness and appropriateness of each is dictated by the predominant culture, resistance to change, magnitude of the performance gaps, etc. (Emiliani et al. 2002).

and every kaizen event. Given that the events are scheduled well into the future and that each of the aforementioned meetings are governed by standard work and established takt times, the lean leader should have virtually no excuses for not fulfilling his kaizen event duties.

COMMUNICATION
Communicating with Team Members

Team members are most effective when, like any employee, they understand at least the most basic elements of their new assignments. These include, but are not limited to the following:

- kaizen scope and targets and how they fit into the "big picture" (that is, the value stream improvement plan);
- timing and location;
- schedule/hours for each day—the day one kick-off time and final report-out should be specified (the remainder of the schedule may reflect approximate start and stop times by day);
- 10–12+ hour days are not uncommon and should be expected;
- roles and responsibilities;
- management expectations (participants are assigned exclusively to the kaizen, clear the calendar of all other commitments, etc.);
- dress code;
- food arrangements; and
- required pre-work, if any.

The team leader and/or the kaizen-relevant value stream or functional manager should contact team members as soon as possible, preferably several weeks in advance. This will enable each team member to make business and personal schedule accommodations. It is unreasonable, counterproductive, and disrespectful to presume that team members can be notified of their assignment to a kaizen team only a day or two before the kaizen itself. Further, it is presumed that the team member's participation has been cleared and communicated with the member's immediate supervisor prior to any initial notification.

Team members who are "surprised" by their assignment are generally unimpressed with management's lack of foresight and courtesy. They understandably become skeptical of the company's competency and credibility relative to any professed commitment to lean transformation and do not soon forget how the company did not value their own imminent work or personal plans. In these situations, team members are less inclined to embrace and drive change or to contribute the effort and long hours routinely needed for a successful kaizen.

The communication, depending on geography, is best conducted by a brief meeting with the entire team and should include, at a minimum, review of the relevant value stream or process maps and improvement plans, copies of the profile, roles and responsibilities (Table 5-1), and kaizen event schedule at-a-glance (Table 5-2), followed by questions and answers. This type of meeting is most beneficial for those who have never participated in a kaizen event. However, it is not necessary for those who are kaizen veterans nor is it practical for those who are out of town or otherwise not available. In such cases, an e-mail (or hard copy) reflecting the profile sheet, roles and responsibilities, and kaizen event schedule at-a-glance, possibly supplemented with a telephone conversation, should suffice.

Similar communications should be made to senior managers and other select personnel. Their attendance is required at the kick-off meeting, daily team leader meetings, and the final kaizen report-out. Presuming that there is sufficient long-range kaizen event planning, it makes sense for the kaizen promotion officer to formally schedule management for the specific critical meetings and report-out. This is much more effective and pragmatic than a "blanket" schedule for the entire week of the kaizen. This approach helps eliminate any excuses for lack of attendance.

The kaizen event schedule at-a-glance (Table 5-2) is a simple and effective way of communicating the timing and location of virtually all elements of the kaizen. The schedule is often supplemented with an e-mail or hardcopy memo (see Figure 5-11 for an example).

Site-wide and Site-specific Communication

Kaizen is about change (for the better) and these changes will, sooner or later, impact all of

Table 5-2. Example kaizen event schedule-at-a-glance

Date	Time	Activity	Location
1/29	6:00 a.m.–6:25 a.m.	Team member introductions and quick review of kaizen area	Team assigned break-out room and gemba (see event kick-off notice)
	6:25 a.m.–7:00 a.m.	Opening comments by plant manager (5 minutes TT), kaizen kick-off presentation by team leaders (10 minutes TT per team)	Second floor training room
	7:00 a.m.–7:10 a.m.	Kaizen kick-off comments and directions (including initial strategy) by consultant to teams and leadership	Second floor training room
	7:10 a.m.–11:00 a.m.	Kaizen training	Second floor training room
	11:00 a.m.–11:30 a.m.	Lunch (provided)	Second floor training room
	11:30 p.m.–4:00 p.m.*	Begin observations	Gemba
1/30	6:00 a.m.–?	Finish observations	Gemba
	11:00 a.m.–11:30 a.m.	Lunch (provided)	Second floor training room
	2:00 p.m.–3:00 p.m.	Team leader meeting (15 minutes TT per team)	Second floor training room
	?–4:00 p.m.*	Define/implement countermeasures	Gemba and team assigned break-out room
1/31	6:00 a.m.–11:00 a.m.*	Define/implement countermeasures, validate new standard work	Gemba and team assigned break-out room
	11 a.m.–11:30 a.m.	Lunch (provided)	Second floor training room
	2:00 p.m.–3:00 p.m.	Team leader meeting (15 minutes TT per team)	To be announced by facilitator
	?–4:00 p.m.*	Define/implement countermeasures, validate new standard work	Gemba and team assigned break-out room
2/1	6:00 a.m.–4:00 p.m.*	Define/implement countermeasures, validate new standard work	Gemba and team assigned break-out room
	11:00 a.m.–11:30 a.m.	Lunch (provided)	Second floor training room
	2:00 p.m.–3:00 p.m.	Define/implement countermeasures, validate new standard work	Gemba and team assigned break-out room
	?–4:00 p.m.*	Wrap up countermeasures, finalize/train to standard work, etc.	Gemba and team assigned break-out room
2/2	6:00 a.m.–8:30 a.m.	Prepare for kaizen final presentation (20 minutes TT)	Team assigned break-out room
	8:30 a.m.–10:00 a.m.		
	10:00 a.m.	Cut-off to notify facilitator if lunch is required for a team working through the day Friday	
	10:00 a.m.–11:10 a.m.	Kaizen event final presentations (20 minutes TT per team)	Second floor training room
	11:10 a.m.–11:15 a.m.	Consultant comments	Second floor training room
	11:15 a.m.–11:20 a.m.	Plant manager comments	Second floor training room
	11:20 a.m.–11:25 a.m.	Awards presentation	Second floor training room
	11:25 a.m.–12:00 p.m.	Event follow-through planning session	Second floor training room

*Teams may stay and work longer if required and desired.
Gemba is the area of the event's focus (Japanese translation is "the actual place"). TT is takt time.

> **Memo**
>
> Date: 2/2/YY
> To: Local management (list names), all kaizen team participants
> From: Value stream manager or kaizen promotion officer
> cc: Relevant senior management/champions, facilitator, consultant/sensei
> Re: Kaizen event 3/9/YY through 3/13/YY
>
> Our next scheduled kaizen events are 3/9 to 3/13 and will be facilitated by _____ with _____ serving as our sensei.
>
> All kaizen team participants, the plant manager, shift supervisors, and department supervisors must plan on attending the kaizen event kick-off presentation at 7:00 a.m. on Monday, March 9 in the training room. The kick-off presentations will progress as follows:
>
> 1. Mr. _____, plant manager
>   Welcome, introduction of guest and kaizen events (5 minutes takt time)
> 2. Ms. _____, team leader #1, _____ event
>   Review of kaizen event area profile and kaizen target sheet (10 minutes takt time)
> 3. Mr. _____, team leader #2, _____ event
>   Review of kaizen event area profile and kaizen target sheet (10 minutes takt time)
> 4. Mr. _____, team leader #3, _____ event
>   Review of kaizen event area profile and kaizen target sheet (10 minutes takt time)
> 5. Mr. _____, sensei
>   Comments and direction on scope, initial strategy, approach, and any immediate training needs
>
> Team members need to be prepared to work from 7:00 a.m. to at least 5:00 p.m. each day. If the team(s) determines it is required, additional hours may be requested. Team members are responsible for coordinating with their kaizen team leader to determine if the team is working overtime on a particular day.
>
> The plant manager, shift supervisors, and department supervisors must plan on attending the daily (Monday through Thursday) team leader meetings. The time, duration, and location of these meetings are noted on the attached kaizen event schedule at-a-glance. During the team leader meetings, the team leader and co-leader from each team will be reviewing the accomplishments made that day, their plan for the following day, as well as any barriers that they have encountered. This is the forum to ask questions, make suggestions, and assist in the elimination of barriers (for example, providing/re-allocating necessary resources).
>
> The kaizen event final presentations (report-outs) will be held on Friday, March 13 in the training room. All kaizen team participants, the plant manager, shift supervisors, and department supervisors must attend. Any other associate whose attendance would be beneficial should be invited. The final presentation is not a forum for questions; it is an opportunity for the teams to communicate their accomplishments and the methodology employed.
>
> Continental breakfast and lunch will be provided Monday through Friday as reflected in the kaizen event schedule at-a-glance.
>
> Kaizen events require a significant amount of hands-on work. Please dress appropriately. This includes the proper personal protective equipment (safety glasses, safety shoes, hearing protection).
>
> Attachments:
> Kaizen event schedule at-a-glance
> Kaizen event area profiles
> Kaizen event target sheets

Figure 5-11. Example kaizen communication.

the employees within a particular value stream. Accordingly, it makes sense to be proactive and give all of the employees within the value stream a "heads-up." They too need to know the details of the forthcoming kaizen event.

Site-wide communication can come in many forms—memos, e-mails, newsletters, department meetings, site-wide meetings, training sessions, etc. The intensity, number, and types of communication modes should be modulated based upon a number of factors, which include:

- the site's lean transformation maturity (for example, is this the first kaizen or another in a long line of successful kaizens?);
- social or cultural milieu (concern over job loss, degree of resistance to change, other recently failed initiatives); and
- anticipated immediate impact on a group of workers—"kaizen affected" (redefined roles, worker movement and need for skills training, redeployment to a kaizen pool or lean promotion office, etc.).

Other than the team members and the (value stream) management team, the most important groups of people are those working in the support functions. They will be providing many of the necessary services (maintenance, facilities, IT, etc.) and thus will be "kaizen affected." Accordingly, these groups should typically receive a more direct and rigorous level of communication and training. A good communications rule is that the more directly someone is impacted by a kaizen event, the more personal the communication should be.

The truly kaizen-affected people represent those who will have to "live" with the changes made by the kaizen team. Clearly, due to team size constraints and composition criteria, many of those affected will not be participants in the kaizen. It is likely, however, that they will be:

- observed during the many times when the kaizen team visits the gemba,
- conferred with when improvements are brainstormed,
- asked to assist in the "trystorming" mode, and
- ultimately, expected to work to the newly developed standard work.

Communication is especially critical in multi-shift operations. Imagine the shock of people who come to work on the third shift only to find that their work area has been rearranged, the work sequence changed, etc. Therefore, those who are kaizen affected should be provided, at a minimum, with lean/kaizen overview training, explanation of proof of the need, and articulated expectations. For example, "You will work in accordance to standard work. If you have a better way that can be supported by data, it will be incorporated into new standard work using the appropriate tools and techniques."

PRE-WORK
Kaizen Event Target Sheet

The profile's "Preliminary SMART Objectives" field provides the feedstock for the kaizen event target sheet. The kaizen team leader, with the assistance of the kaizen promotion officer, should transfer the objectives onto the target sheet (see Figure 5-12).

If appropriate, the team leader should translate the objectives into even "smarter" (specific, measurable, actionable, relevant, and time-bounded) language. This is an exercise in

No fashion sense. It is helpful for people at the gemba to be able to discern who is a kaizen team member and who is just engaged in some "industrial tourism." One practice is to communicate via visual controls—such as the attire of people on the team—wearing distinguishing shirts, hats, or vests. There may have been a cultural subtlety (American versus Japanese) that the kaizen promotion officer failed to pick up on, but after reading a fine lean book, he went forth and purchased a number of red vests and tricked them out with a semi-discreet kaizen logo. At the next kaizen event, he informed the team members of the new wardrobe addition. The teams were largely comprised of women who seemingly were good at recognizing ugly clothing. No one wore the vests. Some months later, the kaizen promotion officer tried to give them away. There were no takers.

Kaizen Event Target Sheet

Location: Aurora
Event: Disbursements
Team lead: John Adams

Team Name/#: AV-#04-6/YY

Date: Apr. 23 – Apr. 27

#	Key Measurement	Start	Target	Day 1	Day 2	Day 3	Day 4	End	IMPACT Difference	% Improvement
1	Existence of A/P processing standard work (Y/N)	N	Y							
2	Reduce lead time from match to voucher (days)	4.5	2							
3	Improve productivity (#invoices/person/hour)	14	25							
4	Reduce A/P process risk priority number (RPN)	TBD	50%							

Remarks:

Figure 5-12. Kaizen event target sheet. (For a blank form, see Appendix A.)

clarity and precision. However, unless there is explicit approval from the executive sponsor, the spirit and the quantified nature of the objectives should not be substantively modified. The targets reflected on the target sheet should have no ambiguity. This is what the team, barring any senior leadership approved mid-course corrections due to scope modifications, better insight, etc., will be trying to accomplish "come hell or high water."

The SMART key driver measure language should be dropped into the second column of the target sheet, "Key Measurement." Preferably, the first word in each entry should be a directional/action verb, for example: "Reduce . . .," "Improve . . .," "Compress . . .," followed by the measure, such as "cycle time" or "lead time" and then, parenthetically, the unit of measure, "(seconds)" or "(units)," etc.

If, for example, one of the kaizen targets is to establish something, such as standard work, then the measurement may read more like, "Existence of standard work." In such a situation, the corresponding target will be digital in nature.

Some target sheet formats reflect pre-populated standard measures. These measures may include:

- space (square feet, square meters);
- inventory (pieces);
- walking distance (feet, meters);
- parts transport distance (feet, meters);
- throughput time (minutes);
- cycle time (seconds or minutes);
- volume per day (units);
- productivity (units/person/hour);
- schedule attainment (%);
- changeover (seconds or minutes);
- quality defects (units);
- quality improvements (#);
- 5S score (points);
- safety improvements (#); and
- energy-saving improvements (#).

Target sheets with a menu of measures can be helpful from the perspective of providing people with possible measures. However, these sheets present two risks: 1) they do not reflect the *priority* of the measures (unlike how it can and should be done with the "blank" target sheets—the first entries should always be the ones reflected on the event profile sheet) and 2) they can facilitate the selection of too many and/or irrelevant measures. Human nature is such that if a measure is on the form, the team leader will likely feel compelled to use it. The deepest thinking relative to objectives and targets selection should occur at the profile preparation stage.

The respective targets should be entered into the fourth column, "Target," while the start or current state or condition should be entered into the third column, "Start." Both the "Start" and "Target" fields should be quantitative as reflected in the parenthetical entry in the "Key Measurement" column. The quantitative terms can encompass time (seconds, minutes, fractions of days, etc.), unit counts (work in process [WIP], steps, number of visual controls, etc.), distance (feet, meters), space (square footage, square meters), or percentages and ratios (units/man-hours, every part every interval, etc.). Sometimes the target is digital from the standpoint that the key measurement reflects establishing something that currently does not exist, for example standard work. In such an instance, the "Target" may be "Yes," "Y," or "All" or "100%," compared to a "Start" of "No," "N," or "N/A" or "0%."

Occasionally, the "Start" is unknown or is not known with certainty. Often, these values will be determined during the kaizen as the team observes reality. However, sometimes, such as in the case of quality oriented kaizens or when

Green targets. While kaizen targets typically focus on big hits, as long as it is not defocusing, targets can be "greener," such as reducing energy use. For example, during a kaizen within a check processing facility, it was determined that the standard work should include turning off the check sorters immediately after the last check had been processed. The result was a greener process, translating into $25k in annual savings.

Tale shared by Michael O'Connor, Ph.D.

addressing processes with extremely long cycle times, the measurement process may require many hours or days of tracking and analysis to derive meaningful data. In this situation, it is often prudent to determine the baseline as part of the pre-work. Otherwise, there is the risk that the kaizen team will expend significant time and effort compiling the baseline data—time that would be better spent on improvements.

The remaining columns of the target sheet are to be completed during the kaizen event, in sequence, just prior to the late afternoon standard team leader meeting. The team leader, as a rule, should be the only one who updates the target sheet. He or she needs to "own" this document. It is essentially the North Star of the event, the most important reference during the daily team leader meeting and the "money" slide in the report-out.

Occasionally, there are situations when the kaizen target sheet format is not the most effective manner to articulate the event objectives. When teams are tasked with developing a new or radically re-engineered process, it is often helpful to use a team mission statement. These are situations in which the current state is:

- highly complex,
- not governed by any modicum of standard work, and/or
- relatively unknown by many from a comprehensive, data-based, direct observation perspective.

A team mission is more prescriptive in nature than target sheets, especially in terms of the de-

> **Too Many Targets**
>
> Consistent with lean and the focus inherent in strategy deployment, the number of kaizen targets must be kept to a manageable few, typically no more than five primary targets. This is not to say other "collateral" objectives may not be enjoyed. For example, while primary targets may include certain productivity improvements, there may be resultant floor-space reductions that can be claimed but were not part of the initial primary targets. Keep the team focused, do not blind them with a large and diverse set of targets.

sired characteristics of the future state, while also re-emphasizing the scope. These elements are critical for a team attempting an improvement that is not a simple derivative from the current state, but more of a wholesale re-engineering of a substantial process. Absent the clarity that a mission can provide, a team can spend a lot of time "spinning its wheels." See Table 5-3 for further insight.

Perceived Barriers

Perception is reality, especially in the context of change management. Perceived barriers, real or imagined, must be addressed to clear the way for effective events.

Perceived barriers should be discussed during the pre-event planning stages of profile preparation and at team orientation activities. Barriers can be technical and/or cultural in nature. Kaizen event planners must have a keen awareness of existing and newly developing risks. They should be tapped into the company "grapevine" and have insight into things that might have imminent operational impact and thus the ability to disrupt the kaizen (for example, a near-term influx of orders).

The goal must be to remove significant barriers or avert anticipated barriers before the kaizen. This requires early identification or forecasting of the barrier, followed by the formulation of a solution or countermeasure and its execution. See Table 5-4 for some examples of possible barriers and approaches for resolution.

Barriers can be diverse and include things like social or workforce issues, schedule or production constraints, and regulatory constraints. Occasionally, new barriers that cannot be readily dispatched before the event must be addressed by the team. Those beyond the capacity of the team due to skill, resources, and/or authority should be elevated to senior management. This is typically done during the event within the context of the daily team leader meeting.

Other Pre-work

The kaizen event is the time and place for the team to focus on a specific scope with the express intent of successfully achieving the targets using

Table 5-3. Formulating the kaizen event mission

Criteria	Example Mission

A mission statement is an appropriate alternative to the event target sheet when:
 	
1. The team must conceptualize and develop a new or dramatically improved process.
2. The current-state has:
   • high complexity (that is, technical complexity, multiple decision points, features, and options), and/or
   • little or no existing standard work.
3. The stakeholders have limited data-based insight into the current state relative to lead time, cycle times, work content, yield, etc.
 	
Event mission must be SMART:
• **S**pecific relative to: 1) scope and 2) future-state characteristics (including lean systems and tools), and performance levels;
• **M**easurable;
• **A**ctionable and **A**ttainable by the kaizen team;
• **R**elevant in terms of leverage and linkage to strategy deployment and improvement plan objectives; and
• **T**ime-bounded relative to what will be accomplished by the end of the event and when future-state plans, if any, will be achieved.
	By noon Friday . . .
	
	Develop a proposal management future-state process map (request for proposal [RFP] receipt through proposal submittal) with:
	
	• one or more high-impact improvements validated through simulation and that incorporate standard work and supporting visual controls; and
	• a process-improvement plan with a projected completion date of 6/1/YY that, among other things:
	
	  –drives substantially improved performance,
	  –50+% lead time reduction,
	  –25+% work content reduction,
	  –100% on time,
	  –clarifies roles and responsibilities, and
	  –incorporates a visual management process that clearly identifies request for quotation (RFQ)/bid status throughout the process and flags abnormal conditions.

Table 5-4. Example perceived barriers

Perceived Barriers	Possible Resolution
Event activities will interrupt operations and fulfillment of customer requirements.	Schedule overtime before, during, or after the event to meet anticipated demand.
Maintenance support will not be sufficient to accommodate anticipated countermeasures.	Secure contract maintenance during the middle days of the kaizen event.
Workers are concerned that they will implement productivity improvements that will eliminate jobs.	Management must communicate location-wide that no one will lose employment as a result of productivity improvements. Within this communication, management must articulate the difference between employment security and job security.

time-proven tools and techniques for identifying, acknowledging, and eliminating waste. It is patronizing, to say the least, to "empower" a kaizen team and then feed them previously compiled data and/or predetermined countermeasures. This is categorically counter to the kaizen philosophy and the inherent respect for the worker. It also falsely eliminates the need for the team to go to the gemba and observe reality firsthand. Pre-work does not include designing solutions! The kaizen process must facilitate a team journey, albeit over a short period of time and one that is often facilitated by a sensei and directed (not managed) by leadership. Within that construct, the journey must enable and allow the team members to substantively identify and acknowledge the waste and related opportunities on their own, and then develop and implement countermeasures.

As reflected in Table 5-5, there are a few acceptable additions to pre-work activities. These exceptions to the rule address specific situations where direct observation of the pre-kaizen situation and/or the necessary rigorous compilation and analysis of historical data would require the team to invest so much time during the kaizen that there would be precious little time to identify countermeasures, "trystorm," and implement and validate improvements. In a similar vein, it is acceptable to conduct pre-event training if it is done no more than a week or so before the event.

LOGISTICS

Poorly executed event logistics will quickly stop a kaizen team in its tracks. Insufficient supplies, missing forms, late lunches, "shifting" break-out rooms, balky projectors, unreachable support people, inaccessible petty cash, and the like demoralize team members, suspend their bias for action, and delay the completion of countermeasures during the events.

It is antithetical to lean when frustrated team members suffer the waste of waiting . . . for an associate to return from the store with newly purchased stopwatches because none were secured before the event. Accordingly, event logistics must be planned and verified prior to the event. While the summary level pre-event planning checklist (Figure 5-3), serves as an excellent logistics planning work aid, it is helpful to use other tools such as a kaizen team supply list (see Table 5-6). For industries that require the movement, storage, and presentation of larger physical items, the kaizen supply list often includes materials such as tubular steel, PVC pipe, and the like, to custom-build things like flow racks, right-sized workstations, and "supermarkets." In any case, team dedicated, visually controlled supply storage cabinets are also required. The generation of a kaizen event schedule at-a-glance also serves as an effective prompt for securing necessary rooms, scheduling refreshments, as well as obviously laying out the sequence of activities and communicating it to the necessary parties.

Table 5-5. Acceptable pre-work

Pre-work	Comments
Documenting reality for operations or transactions with extremely long cycle times	Documenting reality is the critical first step of every kaizen. If an operation/transaction exceeds 8 hours, it may be prudent to document it prior to the event. However, the documentation must properly employ time observation forms, standard work sheets, standard work combination sheets, etc.
Documenting reality for operations or transactions that occur infrequently or are not scheduled during the kaizen	See above.
Compiling historical data for practical, graphical, and/or analytical analysis	This activity may save precious hours and help the team filter root causes for things like unplanned downtime, thereby allowing it to deploy relevant tools such as autonomous maintenance.
Conducting pre-event training for the team(s)	This is an excellent strategy if a qualified person can deliver the training of approved module(s) to the team just prior (no more than a week) to the event.

Gemba Tales

Fresh is better than pre-packaged. During the event kick-off meeting a team leader spent 5 minutes, computer-aided design (CAD) drawing in hand, detailing the pieces of equipment that would be moved out of the cell. This was supposedly to address the team's targets of improving the customer fill rate with a 50% productivity improvement and a 50% reduction in packaging defects. The sensei thought their "plan" seemed a bit pre-packaged and accompanied the team members to the gemba to observe reality.

The target area was a packaging cell comprised of three identical chaku-chaku (load only) stations for low-volume/fast cycle time/high-mix component parts. Recent and planned transfers of new product lines to the plant were driving a need for space savings. The event pre-work included a time observation that indicated one machine was more than capable of meeting daily demand on one shift with one operator with waterspider support. However, over the last month, the cell was manned with six operators over three shifts and still had a substandard fill rate and many defects!

When the sensei and team arrived at the cell, one machine was running, one operator and machine were waiting for parts, and the third cell was down with the waterspider "assisting the operator" with a machine adjustment to get it working. Within 10 minutes, the one running machine was ready for a changeover and became idle. Both idle operators began doing their 5S activities. No supervisors or machine technicians arrived during the next 30 minutes . . . Okay, then . . .

The sensei prohibited the team from removing any machines until the cell demonstrated and sustained the daily demand at the desired fill rates and met the quality performance target on a single shift. In response, the team leader threatened to quit the team citing that they could not execute "their plan" to eliminate two machines and re-layout the kanbans to achieve the space savings they needed for the next cell relocation! Of course, this plan was based upon pre-work that was not consistent with the event targets or the reality that was right in front of them.

The sensei prevailed. By week's end, the cell was producing the daily demand, at the right level of quality, with one operator, and a part-time waterspider, but with three machines! One machine operating, one idle but setup to operate immediately with first good piece completed (one takt time changeover by the operator moving to the machine), and one waiting for changeover or already changed over. To achieve the one takt time changeover with one operator and a part-time waterspider, they reduced space by 40% by re-laying out and resizing the kanbans. A closed-loop overall equipment effectiveness (OEE) improvement plan was created to facilitate the removal of one machine within 30 days if the expected improvements were met.

Morals of the story:

- pre-work is intended to help the team see as much waste as possible so it can create a prioritized plan quickly and eliminate the waste,
- let the data lead the team, and
- a kaizen event is not the forum to execute a pre-established "solution."

Tale shared by Richard A. Jeffrey

SUMMARY

- Pre-event planning is the second phase of kaizen event standard work and should be initiated a minimum of 3 to 4 weeks prior to the actual event date.
- There are four pre-event planning sub-processes:
    1. event selection and definition,
    2. communication,
    3. pre-work, and
    4. logistics.
- Deliverables from the pre-event planning include:
    - kaizen event area profile,
    - kaizen event schedule at-a-glance, and
    - kaizen event target sheet or kaizen event mission.
- Event selection and definition are crystallized through the event area profile development process and then further defined during target sheet or mission statement preparation. The kaizen event area profile

Table 5-6. Kaizen team supply list (For a blank form, see Appendix A)

Pieces/ Units	Item	End of Event Inventory	Quantity Needing Replenishment
Stored within Plastic Storage Bin			
6	Clipboards		
1 set	Laminated copies of standard forms: 1) 5S audit sheet, 2) time observation form, 3) standard work sheet, 4) standard work combination sheet, 5) % load chart, 6) process capacity chart, 7) setup observation analysis sheet, 8) kaizen target sheet, 9) task list, 10) improvement idea form, 11) kaizen newspaper		
6	Stopwatches		
1	Pedometer		
1	25-foot tape measure		
1 box	Pencils (pre-sharpened)		
3	White erasers		
1 box	Pens		
1 box	Flip chart markers (multi-colors)		
1 box	Dry erase markers (multi-colors)		
1	Dry erase eraser		
1 bottle	Dry erase board cleaner		
1	18-in. ruler		
6	8-1/2 in. × 11-in. legal pads		
2	Calculators		
1	Stapler		
2	Rolls of scotch tape in dispenser		
2	Rolls of masking tape		
1 box	Blank 8-1/2-in. × 11-in. overhead projector sheets		
1 box	Paper clips		
1 box	Rubber bands		
3 pkg	Yellow sticky notes 3 in. × 3 in.		
3 pkg	Orange sticky notes 3 in. × 3 in.		
3 pkg	Green sticky notes 3 in. × 3 in.		
1	Scissors		
1 pkg	8-1/2-in. × 11-in. multi-color paper		
1 pkg	11-in. × 17-in. multi-color paper		
1 pkg	8-1/2-in. × 11-in. laminating pouches		
1 pkg	11-in. × 17-in. laminating pouches		
1 box	Sharpies (multi-colored)		
1 box	Push pins		
1	Adjustable 3-hole punch		
Not Stored within Plastic Storage Bin			
3	Flip chart pads		
1 box	Flip chart markers		
Shared Among Teams			
1	Digital camera		
1	Video camera		
1	Label maker		
1	Laminator		
1	Measuring wheel		
1 roll	Kraft paper or white plotter paper		
1	LCD projector (located in presentation room)		
1	Overhead projector (located in presentation room)		
1	Color printer (11-in. × 17-in. capable)		

specifies, among other things: event scope, preliminary objectives, strategy deployment/improvement plan linkage, customer requirements, current situation and problems, and executive sponsor, team leader, and team members.

- Since better teams produce better results, team member selection is an important component of pre-event planning. Guidelines for team member selection include:

  - The typical team size should be six to eight people. Smaller teams often necessitate reduction of the scope of the kaizen and can afford little time to validate improvements. Larger teams are subject to "social loafing" and it is difficult to ensure that all members are performing value-added work.
  - Teams should balance representation to promote diversity, perspective, and ownership.
  - Approximately one third of the team members should be from the target area; one third from up/downstream processes; and the final third should be "fresh eyes."
  - The team leader and co-leader should have the right mix of core and technical competencies.
  - To avoid or minimize bias, it is best if the team leader is not directly responsible for the day-to-day operations of the kaizen target.

- Pre-event communication is a critical element of effective change management. It must be appropriately directed to all of the relevant stakeholders—team members, event support personnel, those within the target process, and the broader value stream and site.

- Care must be taken to ensure that pre-planning and any related pre-work do not design a "solution." These activities are to facilitate and enable kaizen team effectiveness, *not* prescribe specific countermeasures and improvement ideas, which are the responsibility of the team during the execution phase of the event.

- Event logistics may appear trivial in nature, but a team needs timely and appropriate forms, supplies, materials, equipment, meeting space, and refreshments to be effective.

REFERENCES

Emiliani, Bob with Stec, David, Grasso, Lawrence, and Stodder, James. 2002. *Better Thinking, Better Results: Using the Power of Lean as a Total Business Solution*. Kensington, CT: The Center for Lean Business Management, p. 66.

Koenigsaecker, George. 2007. "Sustaining Lean," *Manufacturing Engineering*, May, p. 124.

Stuart, Alix. 2007. "Group Therapy." *CFO* magazine, November. Internet accessed 5-8-09, http://www.cfo.com/article.cfm/10013890/c_2984274/?f=archives.

6
EVENT EXECUTION

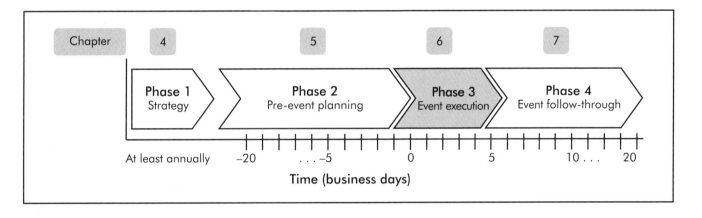

"A good plan implemented today is better than a perfect plan implemented at some unspecified time in the future."
—General George S. Patton

PHASE 3: EVENT EXECUTION

When most people think about kaizen, it is within the context of event execution. The event is unquestionably the more gratifying and action-filled portion of the kaizen process and is markedly different from pre-planning and follow-through (not unlike the distinction between eating dinner and the less glamorous, but necessary, tasks of food preparation and cleanup). That said, the event is much more than the industrial legend of adrenaline-fueled teams executing some sort of mystical plan-do-check-act (PDCA) exercise. Indeed, there is, as in any stable and repeatable process, standard work, inclusive of its own takt time, work sequence and, to a lesser extent, standard work in process.

Although takt time is typically the first element of standard work, the application is a little obtuse when talking about the kaizen event itself. Accordingly, it is easier to first discuss work sequence, followed by standard work-in-process, and then finally, takt time.

Work Sequence

As depicted in Figure 6-1, the kaizen event sequence is comprised of seven basic elements:

1. kick-off meeting,
2. pre-event training,
3. kaizen "storyline,"
4. team leader meeting process,
5. kaizen work strategy,
6. report-out, and
7. recognition and celebration.

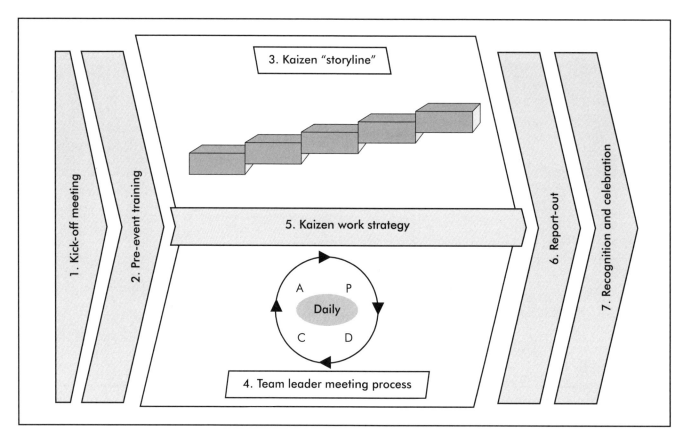

Figure 6-1. Kaizen event sequence.

The event is framed between a formal kick-off meeting and report-out (also known as the final presentation). The kick-off meeting reflects a short, concise launch of the kaizen event in which management, team leaders, and sensei emphasize and share with the team members and others the importance of the event, company support, expectations, kaizen scope, and initial approach. The final report-out represents the brief, obligatory presentation made by each kaizen team to management and peers to show and explain what has been accomplished, including the performance versus the kaizen targets, as well as the means by which the improvements were realized. The content or agenda of the report-out mirrors the kaizen storyline. The final report-out typically transitions to, or is quickly followed by, team recognition and celebratory activities.

The activities between the kick-off and report-out are more dynamic and longer in duration. Pre-event training, perhaps a misnomer because it is typically conducted *after* the kick-off meeting, is exemplary of just-in-time training. It serves as an introduction to the kaizen rookies and a refresher for the veterans, covering the basics of lean and kaizen and, as warranted, the kaizen-specific lean tools and techniques—for example, kanban sizing and transactional process mapping. Pre-event training in no way precludes the formal and informal training that the sensei should introduce throughout the event on an as-needed basis.

The kaizen storyline constitutes the "meat and potatoes" of the event. It embodies the scientific process or journey by which the kaizen team identifies, acknowledges, and eliminates waste and ultimately achieves its targets. On all but the final day of the kaizen event, the team leader, in what is aptly called the team leader's meeting (or team leaders' meeting, if there are multiple teams), provides a short, 10- to 15-minute daily presentation to management, key visitors, and the sensei to apprise them of progress, identify barriers, and seek assistance on barrier removal. Within this same meeting, the team leader(s) also receives guidance, direction, correction, and often exhortation and/or praise from management and the sensei.

Both the storyline and the team leader meeting process intersect in a kaizen work strategy dynamic. The work strategy presents the means for establishing, monitoring, and refreshing the hour-by-hour direction of the kaizen work activities. Assuming solid pre-event planning, the work strategy drives team effectiveness and, ultimately, results.

Standard Work-in-process

Standard work-in-process (WIP), representative of the standard inventory that facilitates the smooth operation of a given process, may not seem relevant to kaizen event execution. However, there are a number of forms and other work products that prescribe and prompt the necessary direct observation, analysis, determination, assignment, and monitoring of action steps, communication of pre- and post-kaizen situations, etc. Within this dynamic, the storyline physically unfolds and provides insight into the kaizen's progress and any related issues and opportunities. Much like any good visual, the standard WIP, depending upon the level of kaizen indoctrination, is self-explaining, team-member managed, and self-correcting.

Takt Time

The event execution takt time is dictated, more conceptually than literally, by the duration of the kaizen event itself. Most kaizen events are 3 to 5 days in length. This represents the numerator of the takt time (T_T) calculation (eq. 6-1), which is typically measured in seconds and sometimes minutes. It is a net number, reflecting gross available time, less any time allotted for things like breaks, lunches, and meetings. For the kaizen, this number is the time that should be specifically dedicated to conducting the kaizen event activities.

$$T_T = \frac{\text{Available time}}{\text{Customer requirements}} \quad (6\text{-}1)$$

It would be presumptuous to carry out a "precise" calculation for the kaizen event available time as the available time is fairly elastic. Leadership and the sensei can exhort the kaizen team members to expand their work hours—"work until you are finished." Of course, length of the workday does not necessarily translate into commensurate outputs. "The forced march" approach is vastly inferior to that employed by a truly engaged team. The engaged team will self-manage to achieve the kaizen targets and, to that end, the mini-milestones (such as the determination of the pre-kaizen situation, identification and prioritization of countermeasures, etc.) as made explicit by the team leader, sensei, or facilitator.

For a kaizen event, the takt time denominator (or requirement) is "1," as in one kaizen event. So, the team has 3 to 5 days, as scheduled, with the option, ability, and obligation to expand daily work time to achieve the kaizen targets. This is one reason why pre-planning is so critical. As in any standard work application, cycle time must be modestly less than (to accommodate work content variation) or equal to takt time. The kaizen event cycle time is driven by the event scope, team size, and swiftness of execution as impacted by things like team selection, initial team strategy, sensei, and leadership support.

Of course, a management time frame of 3 to 5 event days is not very meaningful. You would not dare to kick-off an event and then resurface at the report-out to see if it is "on track." Any good lean person will dissect the event into smaller elements with their own more finite and actionable management time frames. Think of these time frames as the event "pitch," which corresponds to the kaizen storyline milestones. As part of the kaizen work strategy, the experienced team leader will employ almost a day-by-the-hour management plan to track those milestones. This plan should also accommodate the cadence of the kick-off meeting, team leader meetings, and report-out as reflected in Figure 6-2.

KICK-OFF MEETING

The kick-off meeting is the first step within the event sequence. It is the launching point of the kaizen event execution activities. While

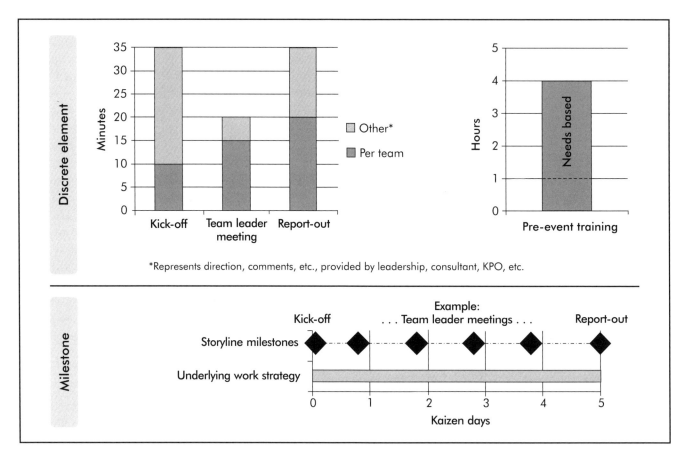

Figure 6-2. Kaizen event takt time(s).

the concept of a kick-off meeting may appear ceremonial in nature, properly done it is extremely value added. See Table 6-1 for an example kick-off meeting agenda.

The meeting, which should be reasonably completed in 30 to 50 minutes, depending upon the number of kaizen teams and excluding training time, is a concise forum in which to communicate, exhort, direct, and teach. The presenters include the kaizen promotion officer, senior and local top management, kaizen team leaders, and sensei. Each team leader's presentation must not exceed a 10-minute takt time. The audience is comprised of kaizen team members, local management and supervisors, and out-of-town guests.

The kick-off provides a forum to clarify important items such as the kaizen week schedule and safety requirements. It also affords managers the "opportunity" (truly an exercise of their responsibility as lean leaders) to express their level of commitment and communicate the absolute importance of the kaizen and its linkage to the company's strategic imperatives. In the kick-off meeting, management is essentially providing and demanding goal clarity while facilitating each team's success through appropriate measures of encouragement, freedom, and challenge.

The kick-off meeting is never the time or place for leadership to question the event selection or scope. This would basically be an advertisement of leadership's lack of competency (they were supposed to figure this out during the pre-planning phase), sow a bunch of discord, and most likely extend the meeting for a long and frustrating time. If there is a substantial last-minute concern over the selection and/or scope, the leaders and several key participants should attend to it immediately after the kick-off meeting.

Possibly the most important kick-off meeting agenda item is the team leader presentation(s).

Chapter 6: Event Execution

Lean Leaders' Short "Must Do, Cannot Fail" List

- Demand and verify that event execution standard work is followed.
- Ensure that lean leaders attend the kick-off meeting and the report-out.
- Ensure that lean leaders attend and participate in the daily team leader meetings.
- Resist the desire to "steer" the kaizen event. Let the data lead the team.
- Communicate and reflect a sense of urgency.
- Require validation of the post-kaizen situation before the report-out.

This provides the first public opportunity for the team leader to clearly and concisely explain the key items reflected in the kaizen event area profile and the kaizen event target sheet: scope, targets, team, takt time, linkage to strategy deployment/improvement plan action items, current process, and issues or problems.

The team leader should also articulate the initial strategy (see Table 6-2 for an example) that the team will deploy immediately after the kick-off meeting (and any training). The initial strategy sets the direction or plan of attack for the team. It first and foremost focuses on the means and methods to conduct the initial observation of reality. The initial strategy is dictated predominantly by:

Table 6-1. Example kaizen kick-off meeting agenda

Presenter: Item	Content	Time (minutes)	Forms/References
1. Kaizen promotion officer: Greetings and orientation	• Welcome everyone • Review kaizen week schedule-at-a-glance • Review safety requirements, facilities, etc.	5	• Kaizen week schedule-at-a-glance • Facilities map
2. Senior/local: Management introduction	• Express commitment to lean and teams • Emphasize need for change, sense of urgency • Articulate expectations for bold action, blame-free environment, etc.	5–10	• Key performance indicator (KPI) trends • Kaizen ground rules
3. Team leader(s): Kaizen scope, targets, initial strategy	• Review event profiles and target sheets • Emphasize linkage to strategy deployment and value stream improvement plans • Outline team's initial strategy	10 per team	• Kaizen event area profile(s) • Kaizen event target sheet(s)
4. Sensei: Opening remarks/direction	• Identify potential team pitfalls/risks • Affirm or suggest modified initial strategy and related training, as required • Encourage teams	5–10	Kaizen storyline
5. Sensei and/or kaizen promotion officer: Refresher or specialized training	Conduct training if pre-event training has not been done, was not sufficient, or did not cover anticipated specific team strategy/technical approach	As required	Various training modules

Table 6-2. Example initial event strategies

Event Type	Typical Initial Strategy
Flow kaizen/value stream mapping	1. Identify product or service family (prepare product family analysis matrix —see Glossary) 2. Lean/value stream mapping (VSM) overview 3. Current state VSM case study 4. Current state mapping of selected product or service family . . .
Standard work—manufacturing	1. Conduct time observations 2. Generate spaghetti charts, time observation forms, standard work combination sheets. . .
Standard work—business process	1. Generate current state process map based upon forensic transaction types that are representative of the process 2. Supplement as appropriate with time observations, spaghetti charts . . .
Setup reduction	1. Record video of operator(s) conducting current state setup 2. As a team, break down the video, populating setup observation analysis work sheet . . .
Material replenishment system implementation	1. Conduct historical demand analysis of targeted items and identify other data pertinent to a plan for every part matrix 2. If production kanban, prepare % load analysis, determine kanban strategy (for example, in-process or batch [signal, pattern or lot-making]), calculate/size kanban

Initial strategies presume that appropriate analysis of customer demand has been conducted as part of pre-planning and that it has been verified and expanded upon during the initial part of the event.

- the kaizen scope,
- targets,
- realistic foreknowledge and understanding (based upon current situation data), and
- the duration of the kaizen event.

The team leader's presentation gives management and the sensei, not to mention the team members (who should have been briefed a week or more previously), near instantaneous insight into three areas:

1. the relevance and the appropriateness of the selected scope,
2. the team leader's level of understanding and forethought as reflected within the initial strategy, and
3. the team leader's commitment and energy level.

Deficiencies in the scope, incontrovertibly the responsibility of local leadership and the kaizen promotion officer, are reflective of pre-event planning process effectiveness and may indicate that the related standard work was not followed. Sooner or later, this may require a re-scoping of the event. Deficiencies in the team leader's understanding may be reflective of a team leader who:

1. was afforded only limited personal preparation time (that is, he or she was surprise "anointed" within the last 24 business hours),
2. does not grasp the situation because he or she is intellectually lazy and did not study/prepare, or
3. simply does not have the technical skill set necessary for understanding what is before the team and how to attack it. Thus the team is at a big risk of at least initially "spinning its wheels" and getting frustrated.

The first situation is due to a lack of adherence to pre-event planning standard work. The last two are due to poor team leader selection, belying a large gap in technical and/or core competencies. Temporary remedies for these types of situations range from close and frequent kaizen promotion officer/sensei support of the team leader throughout the event, to transference of team leader responsibilities

> **Gemba Tales**
>
> **Leave your preconceived solutions at the door.** The team's objective was to reduce late deliveries, a rather important performance measure given that the company was primarily a distributor of chemicals. During the kick-off meeting, the team leader presented the area profile and then stated, "So, what we need to do to fix the problem is get our suppliers to deliver to us on time so we can deliver to our customers on time."
>
> Fortunately, an experienced kaizen promotion officer stepped in and wisely reversed this deadly combination of "cause jumping" and premature "solutions" and got the team to follow the kaizen process. The data-driven process identified suppliers as a lesser issue. It ended up fifth on the list of countermeasures and was appropriately postponed for a future event. In contrast, the top four root causes of late deliveries were internally driven. The team's implemented improvement ideas yielded an on-time delivery performance of 95%, up from 88%. Have faith in the rigor of the process and trust the data.
>
> *Tale shared by John A. Rizzo*

to the co-team leader, to possible assignment of a new team leader.

In the situation of intellectual laziness, it may be useful to conduct a small private post-kick-off meeting to shock the team leader with explicit descriptions of what is expected from him and how the pending daily team leader meetings could be "uncomfortable." Low levels of team leader commitment and energy level are also indicative of a team leader selection shortfall and may require the aforementioned "pep talk" or simply the selection of another, more willing and committed team leader.

The sensei, who will most likely be asking probing questions of the team leaders during their presentations, should have some foreknowledge of the planned event. A good kaizen promotion officer will ensure that the sensei will see at least preliminary copies of the pre-event plan and solicit feedback relative to scope, targets, etc. from him during the pre-event planning phase. The sensei will typically make suggestions to the team leaders during the kick-off meeting relative to initial strategies and potential pitfalls (relative to the observation of the pre-kaizen situation, need for greater understanding of customer demand patterns, etc.). Often the sensei may remind the audience of improvement opportunities that were identified in prior company kaizens (for example, maintenance support was limited due to late communication of team needs to the maintenance department) with the intent that the prior mistakes will not be repeated.

Ultimately, the kick-off meeting will most likely break-off into a training session or sessions. If this is not the case, the teams will be dismissed to their pre-assigned breakout rooms to reflect on the direction/suggestions provided by leadership, the kaizen promotion officer, and sensei, and to gather their thoughts regarding strategy and then attack the gemba.

PRE-EVENT TRAINING

Pre-event training is often seen by leadership as something to just "get through" before really starting the kaizen (that is, doing stuff). However, properly prepared and delivered training provides team members, who many times are apprehensive because of their lack of lean knowledge and limited kaizen experience, with necessary exposure to what they need to know for the event. It also serves as a strong signal that leaders are committed enough to invest time and resources to develop their people and thoughtfully attack lean.

For pre-event training to be effective, its delivery must be:

- well timed and coordinated,
- designed with the adult learner in mind,
- suitable in duration, and

> **What Right Does the Sensei have to "Pontificate?"**
>
> Mostly, senseis have earned the right to pontificate because they have been through hundreds of kaizen events and numerous lean implementations. They are there to teach willing students how to progress farther faster.

- reflective of a curriculum with a breadth and depth designed to satisfy the company's lean transformation needs.

Timing

Typically, training is provided immediately subsequent to the kick-off meeting. However, if the kaizen promotion officer or another individual is competent and the time, team members, and a training facility are available, training may be conducted, in full or part, several days prior to the event. This strategy will enable the teams to spend more time on activities during the kaizen event. Nevertheless, training should be delivered as real-time as possible, approaching just in time to its direct application at the gemba.

Adult Learning

As in any training or knowledge transfer activity, the longer the period between introduction to new material and its actual application, the less likely the trainee will retain the new knowledge . . . "Use it or lose it." This notion of "action" addresses the last of the four key adult learning principles, the first three of which are (Stolovitch and Keeps 2006):

1. readiness,
2. experience, and
3. autonomy.

Readiness reflects the openness, or possibly closed-mindedness, of the adult learner within the learning situation. This receptivity is really dictated by whether the training addresses the participant's needs, for example the solving of a problem, the presentation of an opportunity, or an increase in status and/or professional or personal growth. The groundwork for readiness is prepared by the pre-event planning and related communication and, at a broader level, by the effectiveness of the lean transformation/change management effort. Done properly, employee engagement, and thus readiness, will be exceptional.

Effective training takes into account the level and type of learners' experience. This includes the proper acknowledgement of their depth and diversity of experience, whether it includes lean expertise or not, as well as their perspectives. Training content and delivery should be adjusted appropriately. For example, learners who are advanced on the path of lean maturity should be instructed differently, in manner and content, than those who are new to the subject matter. In the situation where there are varied levels of experience, the more seasoned individuals should be tapped formally (presenting modules or portions of modules) or informally (by means of anecdotal stories) to share their knowledge and experience with others. This teaching responsibility is consistent with the principles of lean and the supporting notion of mentorship.

The pre-event trainer must be mindful of the perspectives and biases of the learners. For example, if a prior lean launch "misfire" presents a barrier to engagement, it may be worthwhile to invest in an upfront team reflection of what previously went well and what did not, and how and why this new experience will be different. Similarly, those who are steeped in alternative approaches (for example, six sigma), with little practical experience in lean, often have to be respectfully "re-referenced" relative to terminology and approach.

Kaizen, at a team and personal level, provides a fruitful and dynamic environment for autonomy. This autonomy, within the "loose-tight" balance of action orientation and kaizen standard work (see Figure 6-3), encourages learners to make their own decisions for the purpose of achieving the kaizen targets. Accordingly, the training

---

Hard Copies of Training Materials

It is often beneficial for team members to receive a hard copy of the training materials. This enables them to write down notes during the training and provides reference material during the kaizen event. Many companies, at a minimum, provide team members with a booklet that contains kaizen refresher training material. Refresher training material content typically includes/addresses: company commitment, lean fundamentals, kaizen event sequence, standard operation forms and their use, copies of blank forms, and a glossary of terms.

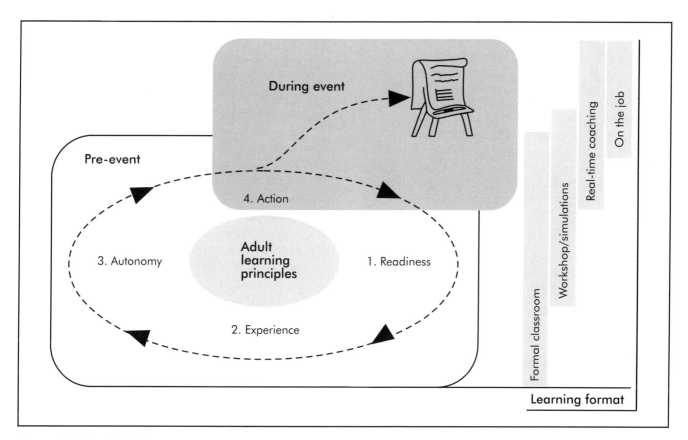

Figure 6-3. Event training.

must respect the capability of the learners, while providing them with as many "reps" (for example, practice in the use of time observation forms) as pragmatically possible. At the same time, the trainer must acknowledge that much of the learning will be on the job and that the process will be coached to ensure team effectiveness, consistency with lean principles, and accountability.

Training Duration and Strategies

Training may be limited to a simple, summary-refresher type class and/or be tailored to specific team needs. The depth and breadth of the training should be adjusted based upon team member experience levels, kaizen scope, and degree of difficulty, cultural readiness, and available time.

It is important to note that pre-event training is *not* a certification or "belt" program. Classroom training should be measured in hours and minutes, not days, and certainly not weeks. Two to four hours is often sufficient to launch a team, especially provided that the participants have been recently exposed to a lean-101-type class. A good coach will deliver supplemental training, formal and informal (flip chart), when needed during the kaizen.

During events comprised of multiple kaizen teams, the teams may be rotated to maximize kaizen time and minimize the likelihood that participants will have to sit in on training that is not relevant to their team scope. A sensei or experienced kaizen promotion officer, with insight into the initial strategies of the kaizen teams and knowledge of the content and duration of each training module, can derive an effective training rotation. For example, assume that the initial strategy of Team 1 includes process mapping, Team 2 setup (reduction) observations, and Team 3 standard work and related observations. The sensei and/or kaizen promotion officer may have the latter two teams go to the gemba to conduct a 5S audit while he provides Team 1

> **Gemba Tales**
>
> **Poor translation.** Pre-event training sometimes provides critical insight into team members' willingness to do their best. During such training an individual, after the sensei asked if anyone knew what the Japanese word *kaizen* stood for, answered "bulls**t." The sensei immediately called for a quick break, wherein he conferred with the kaizen promotion officer and team leader about the crude response. The attitudinally challenged team member was dismissed from his kaizen obligation on the spot. Change management is critical; but spending time on a concrete-head saboteur is *muda*!

with process mapping training. If well timed, the process mapping training will be completed and Team 1 will be dismissed to the break-out room and gemba on or before the return of the latter two teams. Then the sensei may train Teams 2 and 3 on standard work (because they are both relevant and applicable), after which Team 3 will be dismissed to the break-out room and gemba. Subsequently, the consultant will provide Team 2 with setup reduction training.

As previously mentioned, often the sensei or experienced kaizen promotion officer will provide relevant training even more real time. For example, a team working on the development of fixed interval water spider standard work and related material flow may first receive kanban sizing training. Then 24 or 36 hours later, standard work training may be delivered when the team is actually in the position to facilitate replenishment of the kanban by means of a standard water spider route. Similarly, the sensei will make effective use of flip-chart oriented training through which frequent, mini impromptu training sessions are conducted. This method facilitates quick transfer of new tools or concepts to the team.

Training Offerings

A comprehensive lean training curriculum can easily contain dozens of modules. A quick scan of the Lean Certification body of knowledge gives testimony to this fact. The Society of Manufacturing Engineers (SME), the Association for Manufacturing Excellence (AME), and Shingo Prize sponsored Lean Certification defines a specific and comprehensive body of knowledge from which the test questions are derived. The major sections mirror those within the Shingo Prize model; cultural enablers, continuous process improvement, consistent lean enterprise culture, and business results. However, typically team members only need a handful of basic subjects to get started. Good modules on lean (overview), value stream mapping, process mapping, 5S, visual controls, standard work, kaizen, kanban, setup reduction, mistake-proofing and problem-solving can get a team a long way, especially with a good coach who should be capable enough to teach virtually all of it with a flip chart and a trusty marker.

When a company embarks on its lean transformation journey, it usually relies on the outside sensei/consultant to provide the training materials. This is the most expedient and pragmatic approach. However, as a company matures in its understanding and application of lean, the kaizen promotion officer should develop and tailor company "branded" training modules (see Chapter 8).

Curriculum content should satisfy the needs and emphasis of the company's lean business system and the underlying strategic breakthrough objectives and value stream improvement plans (VSIPs). This means that it is predicated upon "pull." In other words, if the VSIP does not call for total productive maintenance (TPM), then that module is probably not needed right now.

The corporate kaizen promotion officer or designees should maintain a formal training module inventory list as well as the revision-controlled electronic training documents themselves. The read-only electronic versions can be made available on the company network for easy access by trainers and self-motivated learners. Companies often use a champion approach by which individual employees serve as lean tool subject matter experts. These champions develop and maintain the training, serve as a trainers of trainers, compile lists of reference materials, etc. Table 6-3 provides an example lean training module inventory list.

Table 6-3. Example lean training module inventory (extract) (For a blank form, see Appendix A)

Module Title	File Name	Description/Purpose	Number of Slides	Training Duration (minutes)	Editor/Content Expert	Last Revision	Presenter Notes (Y/N)
1. Lean introduction	LeanIntro.ppt	Provide overview of lean basic principles and concepts, introduction of lean tools	127	120	O'Connor	6/5/XX	N
2. Value stream analysis (VSA)	VSA.ppt	Introduce value stream concepts and analytical methods, including a case study; use to facilitate value stream analysis in a workshop/kaizen format	52	90 plus time to conduct actual VSA	Horton	1/28/YY	Y
3. Kaizen	Kaizen.ppt	Introduce kaizen event process, storyline, roles, responsibilities, key forms, etc.; use to prepare kaizen team members and management for event	35	60	Lowell	10/15/XX	N
4. Process mapping	ProcMap.ppt	Demonstrate process mapping methodology; distinguishes between value stream analysis and flow charting	41	45	Papelbon	7/21/XX	N
5. 5S and visual controls	5S.Visual.ppt	Introduce 5S (elements and application—audits, red tagging, etc.) and visual controls, including many before and after pictures	79	60	Tito	2/5/YY	N
6. Standard work	StdWork.ppt	Introduce standard work and related elements; exercises require participants to use time observation forms, standard work sheets, and standard work combination sheets; a review of process capacity sheets and operator balance charts is also given	64	120	Lester	5/1/XX	Y

KAIZEN STORYLINE

The kaizen storyline, together with the team leader meeting process and kaizen work strategy, represents the heart of the kaizen event. The storyline, more generically defined as "the plot of a book or play or film," is, in this instance, the plot of the kaizen. It serves as a veritable roadmap for the kaizen team.

Before embarking on a detailed exploration of the storyline, it is worthwhile to compare it, at a high level, to the "Kaizen/QC Story" (Imai 1999) and the six sigma define, measure, analyze, improve, control (DMAIC) methodology (briefly outlined in Chapter 2). This comparison of the three methodologies, along with their respective intersection with plan-do-check-act (PDCA), is reflected in Figure 6-4. It must be noted that because the kaizen event time frame is significantly compressed in comparison to the traditional six-sigma project, the approach is much more aggressive in nature.

While the storyline may appear output oriented in contrast to the more action oriented "Kaizen/QC Story" and DMAIC paths, this characteristic is representative of its dual role: 1) as an outline of outputs with the necessary implicit actions to drive those outputs, and 2) as a secondary mode of dynamic "storyboarding." When distilled down to the level of key documents and forms, especially in a visually controlled environment, this "plot" takes on a character and utility closer to that of a storyboard—"a panel or series of panels on which a set of sketches is arranged depicting consecutively the important changes of scene and action in a series of shots (as for a film, television show, or commercial)" (Merriam-Webster OnLine 2009). This sounds like a perfect medium for 21st century "kaizeners."

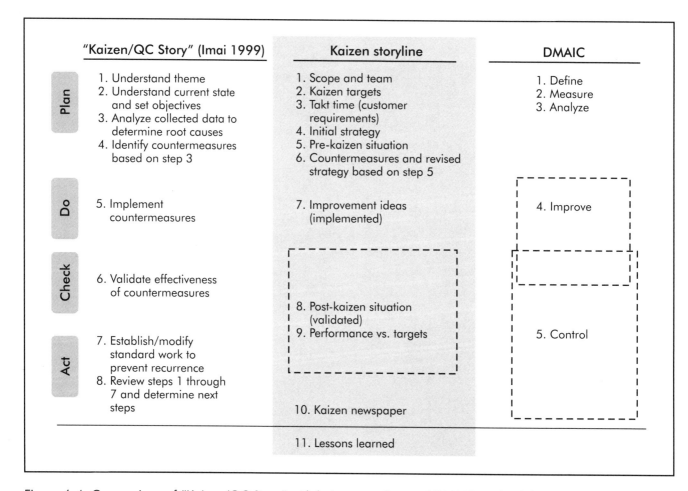

Figure 6-4. Comparison of "Kaizen/QC Story" with kaizen storyline and DMAIC methodologies.

The storyline communicates a message while providing a roadmap or path that begins with the articulation of scope and team (see Figure 6-5). It then progresses, step by step, until the establishment of the post-kaizen situation (which is, of course, validated by direct observation during the event itself), the recognition of the final performance versus target, the identification of the post-kaizen action items (kaizen newspaper), and team reflection on lessons learned.

The storyline contains, at a generic level, the scientific process inherent in kaizen. It shows not only what is to be done, but also how it is to be done, from the perspective of learning, strategizing, and following a process that encompasses observation of reality, brainstorming, "trystorming," verification, quantification, and standardization. The event storyline facilitates the proper use of the various standard kaizen forms, while the forms facilitate a natural flow or progression.

It is extremely helpful, and visual, to tape the related forms, charts, and pictures on the kaizen team's break-out room wall in the proper sequence. By using this technique, the storyline is then truly transformed into a storyboard, making the event progression and logic evident to the kaizen participants and any other observers. This is visual management!

It is important to note that the storyline, as reflected in Table 6-4, begins its definition in the pre-event planning phase as part of the area profile preparation and ends with the event execution phase. The profile, as discussed previously, reflects the kaizen scope, preliminary objectives, team composition, and insight into takt time. The balance of the storyline is rounded out during the event. The initial strategy should be articulated during the kick-off meeting, while the remainder of the storyline is addressed within the body of the event and the wrap-up. This

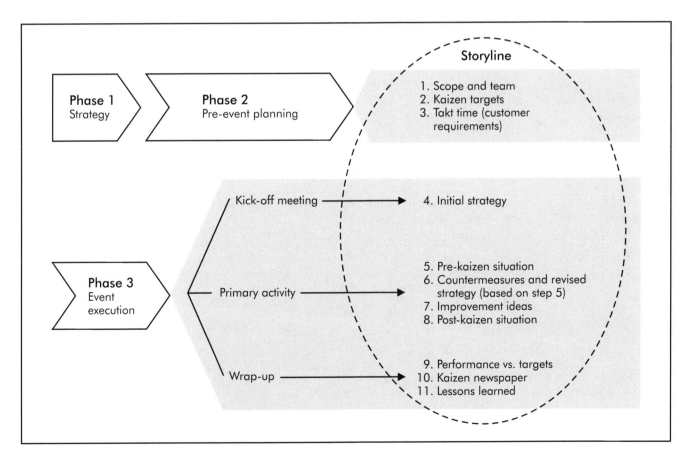

Figure 6-5. Storyline elements by kaizen phase.

Table 6-4. Kaizen storyline

Item	Purpose/Process	Example Tools/Data Forms	Example Tools/Data Other
1. Scope and team	• Provide team with critical background data • Focus team on finite, relevant, business/value stream critical opportunity	Kaizen event area profile	Flow charts, Pareto charts, etc.
2. Kaizen targets	Focus team on specific, measurable, actionable, relevant, and time-bounded performance targets	Kaizen event target sheet	Kaizen mission statement, if applicable
3. Takt time	Quantify and understand rate of customer demand as well as mix and variation	n/a	• Historical demand • Forecasted demand
4. Initial strategy	Formulate/articulate initial team strategy, primarily related to observing reality (pre-kaizen situation)	n/a	n/a
5. Pre-kaizen situation	Observe situation and gather factual evidence to identify, understand, and acknowledge waste and its root causes	• 5S audit • Time observation form • Standard work sheet • Standard work combination sheet • % Load chart • Process capacity chart • Setup observation analysis work sheet, etc.	• Process maps • Historical data • Interviews • Cause-and-effect diagrams • Failure mode and effects analysis (FMEA), etc.
6. Countermeasures/strategy	• Brainstorm and prioritize countermeasures to best eliminate identified/acknowledged waste • Devise strategy to "trystorm," sequence, assign, etc. • Revise scope and targets as appropriate	Task list or countermeasures form	Flip chart
7. Improvement ideas	"Trystorm," simulate, validate, and implement key improvements to achieve kaizen targets	Improvement idea	n/a
8. Post-kaizen situation	• Document and quantify verified improvements • Prove-out, post, and train workers on new standard work	See #5, pre-kaizen situation	• Procedures • Work instructions • Visual controls, etc.
9. Performance vs. kaizen targets	Display performance versus targets	Kaizen event target sheet	Kaizen mission statement, if applicable
10. Kaizen newspaper	Identify and assign open countermeasures	Kaizen newspaper	Project plan
11. Lessons learned	• Promote team member reflection on kaizen experience • Identify things to start, stop, and continue doing	n/a	Flip chart

wrap-up portion, which includes the final (within event) look at validated performance versus kaizen targets, finalization of the kaizen newspaper, team reflection, and capture of lessons learned, is essentially report-out preparation.

Pre-kaizen Situation

The kaizen team's initial strategy, mindful of the ultimate kaizen targets, prescribes the method for gaining insight into the pre-kaizen situation. Specifically, the team seeks objective evidence, through direct observation, if at all possible and pragmatic, to identify the waste within the current condition. To borrow a term from the great Hiroyuki Hirano (and truncate its meaning a bit), the kaizen team must practice "wastology." This is loosely defined as the scientific study of conditions to identify waste and its root causes so that it can be ultimately eliminated. Not to muddy the waters too much, but it must be acknowledged up front that "wastology" can get the team only so far. Beyond waste identification is waste acknowledgement, and beyond waste acknowledgement is the execution of countermeasures to eliminate that waste.

It is important to understand that the "cascading" threefold path of identification, acknowledgement, and elimination is multiplicative, much like a rolled throughput yield calculation. In other words, as reflected in Figure 6-6, a team can only acknowledge what it has identified, and it can only eliminate what it has acknowledged. To make things more challenging, the drivers of each step contain technical and behavioral elements. The long-winded corollary to, "You can lead a horse to water . . ." is "You can teach a person to identify waste, but you can't make him acknowledge it . . . and for what he does acknowledge, you can't always make him aggressively try to eliminate it."

There are a variety of methods and tools for identifying waste. Their effectiveness is largely driven by three characteristics:

1. an explicit requirement of direct observation,
2. the obtaining of specific data, and
3. the recording and displaying of data in a specific format that facilitates analysis, understanding, and discovery.

Flow Kaizen

The event sequence for flow kaizen does not follow as rigorous a path as that employed in a process kaizen. The supporting elements of kick-off meeting, report-out, and team leader meeting are often rendered redundant because the management team and the flow kaizen participants are one and the same. If this is not the case, then there is a distinct possibility that the team has not been properly selected and/or staffed. However, this is not to say that there is no place for a kick-off meeting or report-out. These can provide a venue for senior executives, who are organizationally well "above" the value stream and its direct ownership, to express their commitment and interest in understanding and, thus supporting, the achievement of the future state map by means of the value stream improvement plan.

The flow kaizen storyline differs from that followed in a typical process kaizen. For example, the storyline often encompasses:

1. team and scope (event area profile),
2. lean overview and value stream mapping training with a current state case study,
3. final definition/selection of product or service family (product family analysis matrix, takt time review, etc.),
4. gemba walk,
5. current state map,
6. lean value stream training and future state case study,
7. future state map,
8. strategy deployment cross-check,
9. value stream improvement plan, and
10. lessons learned.

Direct Observation

Direct observation represents the intersection of the gemba with *genchi genbutsu*, Japanese for "go look and see," to gain an understanding of the situation. Taiichi Ohno, the acknowledged father of the Toyota Production System, was famous for the chalk-marked "Ohno circles" that he drew on the plant floor and in which he exhorted his pupils to stand and observe, sometimes for hours on end.

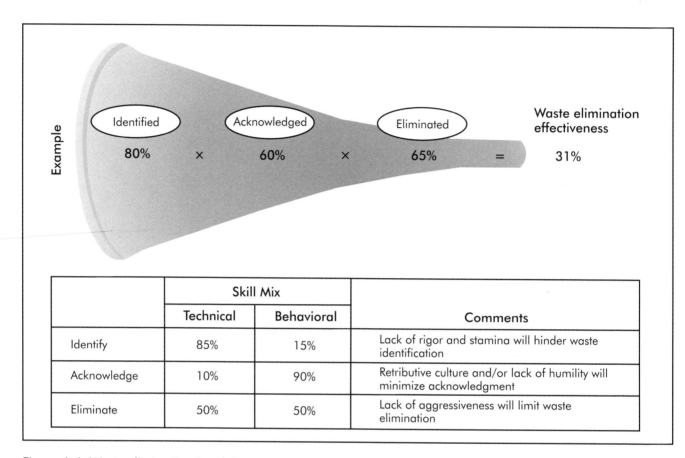

Figure 6-6. Waste elimination "math."

Surely, nothing replaces direct observation as evidenced by the countless executives who, while participating in a kaizen event, express their utter amazement that there is so much waste in the process. Did the waste just recently appear for the first time or was it always there, hidden from the all-knowing, all-seeing executive who never quite had the time to stop and observe a complete cycle within a process in excruciating detail?

Now, what about processes that have cycle times of many hours? Is there an expectation that the process must be directly observed? The answer is usually a painful "Yes." It is impossible to improve what is not understood. This is not to say that the observation time cannot be managed by breaking the process into sub-processes or segments (this should be reflected in the kaizen scope). However, with long cycle times there is a limitation to the number of cycles that can be observed. Pragmatically, sometimes there may be time for only one or two observations. In this situation, the team must understand that the opportunity to witness process variation, a critical element, is essentially zero.

Similarly, there may be times where the process lead time is extremely lengthy. For example, a bodily injury insurance claim, depending upon the complexity, may have a lead time of many months. Is the kaizen team expected to observe this process directly? No! The expectation, however, would be to process map the claim (and others) using the forensic evidence of the claim files as supplemented by interviews with the claims adjuster and others, as necessary. Forensic evidence, while not replacing direct observation, is the next best thing. Think "CSI kaizen."

The ten basic waste identification tools reflected in Table 6-5 represent tried and true ways of identifying waste. Each tool, when properly applied, requires its users to conduct direct observation (the exception being the process map, which

Table 6-5. Ten basic waste identification tools

Common Name (Also Known as)	Description	Typical Application	Insight into Waste
1. Current state value stream map (material and information diagram)	• Visual portrayal of product or service family's material and information flow • Data boxes, lead time ladder, and rolled throughput yield (RTY) line provide summary of process performance and flow issues • "Issue bursts" flag barriers/areas of opportunity	• First high-level step-in development of future state value stream map and related improvement plan • Best in repetitive or transactional scope	• "Issue bursts" capture team identified conditions that typically hinder flow and pull • Examples include push production, cycle times in excess of takt time, unnecessary process steps, etc.
2. Process map (deployment flow diagram)	• Flow chart or map of processes and decisions • Reflects linkages between functions, critical path, lead times, work content, and RTY • Value-added and non-value-added activities differentiated • Current state "issue bursts" flag barriers/areas of opportunity	• Often used to document reality for long lead time administrative processes with many steps • Can use forensic data (for example, files, transactions, memos, etc.) when direct observation is difficult	Map makes evident multiple hand-offs, quality checks, queue time/inactivity
3. 5S audit sheet	• Reflects qualitative scoring criteria for each element of 5S • Determines status/score of each "s" and in total • Provides format to record relevant observations	Useful for developing baseline and periodic status scores for targeted area	• Assists in identification of 5S gaps and opportunities • Identifies situations (for example, non-point-of-use storage, excess tools, materials, improperly labeled references, tooling, etc.) that can drive waste of motion, transportation, defects, etc.
4. Time observation form	• Documents direct observation of processes performed by a single operator • Requires identification/recording of component tasks and task time with use of a stopwatch • Multiple cycle observation facilitates the determination of least waste and/or most repeatable cycle times • Provides field to record "points observed"	• Excellent tool for repetitive processes • Serves as feedstock for standard work sheet, standard work combination sheet, and % load chart	• Forces documentation and quantification of smallest observable tasks (both value added and non-value added) • Provides opportunity for insight into all 7 wastes • Identifies cycle time variation and facilitates the recording of points observed to explain variation and/or note opportunities for improvement

Table 6-5. (continued)

Common Name (Also Known as)	Description	Typical Application	Insight into Waste
5. Standard work sheet	• "Plan view" sketch of operator movement relative to machines/workstations and materials (and virtually through the different computer screens as within a call center) • The form reflects the three elements of standard work: 1) takt time (and net cycle time), 2) work sequence, and 3) standard work in process (WIP) • Also reflected are quality checks and safety devices/precautions • Companion document to standard work combination sheet	Applicable for repetitive processes, but also can be useful for non-repetitive processes in the form of a "spaghetti chart," which traces the physical or virtual path of the operator	Highlights waste of transportation, motion, and excess stock on hand
6. Standard work combination sheet	• Numerical and graphical portrayal of the least waste or most repeatable elemental combination and sequence of a given worker's manual, walk, and wait time, as well as related machine (auto) time • Compares total cycle time versus takt time • Companion document to the standard work sheet	Excellent tool for repetitive processes	• Provides a visual perspective on manual, walk, wait and machine time • Facilitates countermeasure development for elimination of high waste or high work content as well as re-sequencing work for the operator or reassigning tasks to other workers
7. % Load chart (operator balance chart, operator loading diagram)	Graphical comparison of takt time and operator cycle times (and/or elemental cycle times)	Useful in repetitive multi-operator processes	• Provides a visual perspective on load among operators as load imbalance increases the risk of overproduction and/or waiting • Load at or near takt time will facilitate one-piece flow, productivity improvements, and highlight opportunities for work content reduction and redeployment of workers

Table 6-5. (continued)

Common Name (Also Known as)	Description	Typical Application	Insight into Waste
8. Process capacity sheet	Documents the calculation of true machine capacity accounting for machine cycle times, tool setup and change intervals, and manual work time	Used in real or presumed machine capacity constrained conditions	Helps identify opportunities for increasing machine capacity through better staffing/management and continuous improvement (for example, reducing tool change time) rather than purchasing an additional machine
9. Setup observation analysis work sheet	Documents the direct observation of setup activities by operator for the purpose of identifying smallest observable tasks (both value added and non-value added) and categorizing them into internal and external time with the ultimate objective of setup reduction	Any setup	• Facilitates the shifting of internal to external setup time • Facilitates the reduction of internal and external time
10. Operations analysis table	• Numerical and graphical portrayal of process steps • Reflects categorization of each step as processing, material handling, conveyance, idle time, and inspection take place, while quantifying the step's cycle time and distance traveled, if applicable	Excellent tool for repetitive processes	Quickly highlights non-value-added activities and their related magnitude (time and distance)

is typically a forensic exercise). The tools also specify the data that should be captured.

Obtaining Specific Data

One major risk in any data collection effort is the lack of standard work relative to what should be observed, when it should be observed, and at what frequency it should be collected. The ten tools in Table 6-5 provide guidelines for collecting data. For example, absent a time observation form or standard work sheet (spaghetti chart), a kaizen team member will often begin his observation with keen intent and then his attention will quickly diminish or shift to other people or things. It is human nature. However, this same person (recently trained) with a time observation form in hand and working in tandem with another team member on the stopwatch is there with a purpose, recording component tasks and the related cumulative times over multiple cycles and points. This observer is engaged, invested, and is now infinitely more aware of the specifics of the process—the steps, sequence, cycle time, variation, etc.

The foregoing example demonstrates the symbiotic relationship between direct observation and specific data requirements. It not only ensures an effective waste identification process; it also helps develop the kaizen event participants' "eyes for waste." The value of trained eyes, from both a cultural/change management and lean technical perspective, cannot be overestimated.

Recording and Formatting Data

The formatting or distillation of the data obtained through direct observation facilitates analysis at an individual and team level. This enables the team to arrive at an often-obvious discovery and quantification of waste which, in turn, leads to the determination of its root causes. This is really where kaizen, as the flow of ideas, begins its journey.

A classic example of a tool used to distill data is the % load chart, also known as the operator balance chart or operator-loading diagram. The % load chart is constructed by drawing the takt time across the Y-axis and then, in bar chart formation, reflecting the cycle times (derived from the source data—time observation forms and/or the standard work combination sheets)

Gemba Tales

Observing till it hurts. Observing and documenting reality, sometimes in *excruciating* detail, is necessary to remove bias and emotion as barriers to improvement.

Knowing that he was entering into an emotionally charged and somewhat confrontational situation between management and the United Steelworkers, the sensei ensured that the kaizen team (appropriately comprised of workers and managers) was most diligent in its direct observation of the subject operation. This included well over 8 hours of observation, followed by another day of documentation to complete the standard work sheets, standard work combination sheets, and % load charts. When the sensei asked the team members what the data was telling them, the management participants meekly bowed their heads and said nothing knowing that the data clearly indicated that one person, versus the present staffing of two people, could easily perform the job requirements and the other worker would have to be redeployed to another part of the plant. The Steelworker participants were equally mute, but for different reasons. After nearly 30 minutes of this type of probing by the sensei, along with his repeated review of the data, which was hung on the wall for all to see, the local president of the Steelworkers finally burst out, "What do you want me to tell you? This job only needs one person! Alright, this job only needs one person!"

The sensei replied, "If that's what the data is telling you . . ." The team had reached consensus, despite weak leadership, because of the preponderance of data.

Standard Forms = Visual Language

After learning how to read a standard form once, a person can easily apply the new skill. The meaning of complex data is discerned in a glance. It is a language more quickly absorbed and more accurately conveyed. When a whole organization shares this language, the impact is profound.

for each relevant operator. The resulting chart provides, in many instances, a startling visual of the balance or lack of balance between operators. It provides insight into opportunities for work content reduction, re-balancing, and continuous flow. Accordingly, it is the format of the data and/or underlying calculations using that data, which enable the kaizen team to better see the waste and the opportunity.

Symptom or Root Cause?

The simple intent of the pre-kaizen situation is to:

1. identify and analyze symptoms (waste),
2. formulate theories as to the root causes,
3. test the theories, and
4. identify and verify the root causes.

Once these four steps are completed, the countermeasure part of the storyline can be addressed.

Much of the time, the waste identification tools not only facilitate the identification of waste, but they point directly or at least tangentially to its root causes (see Figure 6-7). In other words, skipping from symptoms to identifying the root causes is a "no-brainer." For example, a spaghetti chart that reflects an operator's lengthy "safari" to retrieve supplies, materials, tools, or information provides some good insight into the root cause of the waste of motion, which probably includes the lack of point-of-use storage.

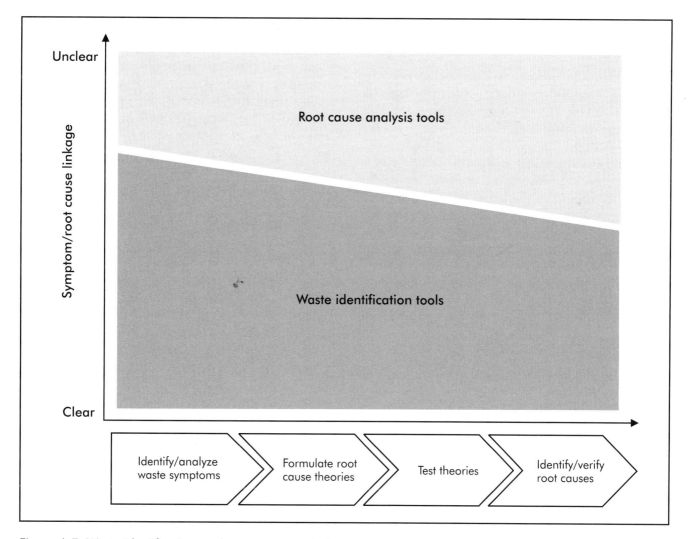

Figure 6-7. Waste identification and root cause analysis.

> **Out of Scope = "Parking Lot"**
>
> Especially during the analysis of the pre-kaizen situation, team members will raise numerous issues or problems. Sometimes these are unrelated to the kaizen scope. Despite that fact, often the team members will dwell on the item(s), and "chase their tails." A good facilitator will quickly identify that the issue or problem is unrelated to the kaizen scope. To get the team to refocus and move on, he will summarize the item in language that specifically captures the concern and record it on a "parking lot." The parking lot, a simple flip chart that captures the problem, enables team members to feel comfortable that the item will not be lost. They can then turn their attention to more relevant issues. The parking lot items should be reviewed with management off-line after the event to determine action plans.

However, it is not always so easy or obvious to verify the root causes of waste. Indeed, if the root cause is not verified, but only the symptom, the countermeasures will be flawed and the waste may not be eliminated. In fact, the team may have created even more waste!

The waste identification tools provide or facilitate, in six sigma parlance, the first two of the three analytical phases: 1) practical analysis (meaning the raw data), and 2) graphical analysis (graphs and charts). While the waste identification tools certainly are not the final answer to the first two phases, they wholly leave pursuit of the third phase, analytical analysis (together with the first two phases—think "P.G.A.") to other, more specialized tools.

Some analytical tools are quite simple and powerful. Others, while being powerful, may need a six-sigma green belt or black belt to facilitate, along with some statistical software. The more complex analytical tools like regression analysis, design of experiments, and hypothesis testing have their time and place.

The simple tools include the 5 whys, cause-and-effect diagrams, and data collection forms (check sheets and concentration diagrams) as supplemented by funneling tools like process failure modes and effects analyses (FMEAs) and graphical analysis like scatter diagrams, histograms, and Pareto charts. These eight basic root cause analysis and supporting tools are summarized in Table 6-6.

As the team converges on the identification of root causes, it is helpful to record each root cause on a separate sticky note. This should be a facilitated process with full team engagement. The language on the sticky notes must be brief, but precise to ensure that there is no commingling of multiple root causes.

Should the team members find that they have a large number of sticky notes, they should engage in an affinity exercise in which sticky notes are physically moved and grouped, whether on a flip chart or a table, into like "themes." A proven method for this exercise is for each team member, one by one, to approach the sticky notes and move them as he or she sees fit into similar groupings. The next team member will do the same, sometimes repositioning what the person before may have done. This may take several cycles through the entire team, but typically a handful of root cause themes or categories will emerge. The facilitator or team leader should gain consensus on the category titles, record them (usually on different color sticky note), and position them above the appropriate groupings. With a finite population of root causes, the team is most probably ready to identify the relevant countermeasures.

Countermeasures/Strategy

At this juncture in the storyline, the team, with a firm understanding of the pre-kaizen situation, is primed to identify and plan the transformative actions that will ultimately create the post-kaizen situation. At the same time, the expanded insight into the pre-kaizen situation, as opposed to what the pre-event planners had prior to the kaizen, may instill a desire to modify the kaizen strategy—including scope and targets.

Countermeasures

Countermeasure is a word frequently used in kaizen. But what exactly does it mean? One definition states it is "Something that is done in reaction to and as a defense against a hostile

Table 6-6. Eight basic root cause analysis and supporting tools

Common Name (Also Known as)	Description	Typical Application	Insight into Root Causes
1. 5 Whys	Form/process that facilitates the asking of 5 whys (and often one "how") to determine root cause of the problem	• Immediately upon identification of a problem/recurring problem • Requires initial problem statement from which the 5 whys start	Ensures deeper penetration beyond primary and secondary symptoms to the root cause of the targeted waste
2. Cause-and-effect diagram (fishbone diagram, Ishikawa diagram)	• Visual portrayal of the potential causes and their relationships (primary, secondary, tertiary, etc.) for a narrowly defined problem • Potential causes are typically categorized by theme (for example, environment, material, machine, person, procedures, etc.)	Used when there are a large number of potential causes and/or there is ambiguity relative to the relationship between the potential causes	• Assists in identification of possible root causes, their relationships, and "measurability" • Hypothesized probable root causes can and should be tested to verify, exclude, and/or determine the magnitude of their contribution to the problem
Data collection forms (basic forms designed to standardize data collection, for example, check sheets, concentration diagrams, frequency plot check sheets, and confirmation check sheets)			
3. Check sheets	• Form on which the characteristics or conditions of interest are reflected • Allows the data gatherer to record frequency as well as comments for each occurrence	Used for situations such as downtime and other process upsets or interruptions	Provides data for what are usually unmeasured (anecdotal) root cause issues or problems
4. Concentration diagrams	Visual reference of location and type of defect or problem as experienced in data entry/analysis (physically marked/coded on the form itself) or in a fabrication, assembly, test or inspection process (physically marked/coded on product drawing in the relevant location)	Used often to identify missing, incomplete, or inaccurate data on forms, screens, etc., and/or damage or defects on a given product	Provides type and occurrence data that will better isolate and direct further root cause analysis
5. Scatter diagrams (scatter plots)	Graph that reflects the relationship between two variables (for example overtime hours and errors)	Use to identify possible relationships between changes observed in two separate sets of variables	Provides insight into cause-and-effect relationship between two variables, although a relationship does not always imply causation
6. Histograms	Graphic summary of variation in a set of data over a given distribution	Excellent tool to see relative frequency of occurrence of various data values	Enables patterns to emerge that are reflective of process capability
7. Pareto charts	Graphic ranking of factors related to problems/issues	Powerful tool to scope/re-scope team on critical few issues	Enables team to identify and focus on the vital few factors that may be root causes or lead to the identification of root causes
8. Process failure modes and effects analysis	• Matrix that reflects process steps and failure modes and effects while ranking risk of severity, occurrence, and (in)ability to detect the failure mode • Calculates individual process step and an overall process risk priority number	• Use when there is a lack of insight into important variables and how they impact quality • Can be used for proactive risk assessment	Useful for identifying important variables within a process that affect/may affect product or service quality

action by somebody else, or something that is done in order to deal with a threat" (Encarta World English Dictionary 2009). The definition sounds militaristic in nature. However, this theme is appropriate in that countermeasures are typically identified by team members for the purpose of eliminating waste (muda), unevenness (mura), and/or unreasonableness (muri). This is, in a sense, war.

The word "countermeasure" is also distinctly different from what the uninitiated may use instead, "solution." *Solution*, at a minimum, infers that a problem is fixed, done ... gone. This is a rarity in the dynamic world of continuous improvement; the "solution" may address the root cause only temporarily, or possibly only the symptom (bad), and/or possibly cause or highlight other issues or opportunities somewhere else in the value stream.

The language of countermeasures is action oriented and in direct response to issues or problems identified by the team. For example, if it has been determined by observation that there is the waste of motion in which the operator must spend time searching for tools, then the countermeasure could be something like, "Establish visually controlled point-of-use storage location of tools (and only those tools) required."

Every countermeasure must be tested to see if it is consistent with lean (for example, a countermeasure that seeks to create an elegant cell to conduct rework rather than addressing the root causes of the defects that necessitate the rework is *not* lean). Countermeasures also must be prioritized; not every countermeasure can and should be executed during the event.

To facilitate the prioritization process, the team can employ an affinity process similar to that described earlier with the sticky notes. The fully engaged team, immediately after the review of the waste and the root causes of the waste within the pre-kaizen situation, should brainstorm countermeasures. Each countermeasure should be recorded on a separate sticky note. Often these countermeasure sticky notes can be physically "matched" to the related root cause(s) sticky notes to visually confirm the linkages. There should not be any unaddressed root causes. An affinity exercise also can be conducted to identify countermeasures that are essentially duplicates (they may use slightly different words) and to identify countermeasure themes, for example, standard work, visual controls, etc. A symbol or code that reflects these themes can be recorded on a corner of the note for later reference. This thematic identification may seem non-value added, but it is often effective to assign countermeasures to team members or groups of team members based on themes. For example, the energy behind a sub-team focusing on visual controls or a material replenishment system can be impressive.

Typically a kaizen team will identify a large number of countermeasures, sometimes upwards of 20 to 30 or more. While some countermeasures are "quick hits," many are not. This translates into a total work content that often exceeds the capacity of the team and those who are supporting it during the event. Accordingly, the countermeasures must be prioritized.

Using the process reflected in Figure 6-8, the full team, as facilitated by the kaizen promotion officer or sensei, should position each countermeasure sticky note on a hand-drawn, four-quadrant prioritization matrix (this also can be done on a wall with the quadrants delineated by masking tape). The quadrants are reflective of a y-axis (impact) and an x-axis (timing). The "northern" half, "A," of the y-axis represents high impact relative to satisfying the kaizen team targets; the "southern" half, "B," represents low impact. Meanwhile, the "western" half, "1," of the x-axis represents countermeasure implementation timing *within* the duration of the kaizen; the "eastern" section, "2," represents countermeasure implementation timing *after* the kaizen event is completed. The four quadrants can be referenced as A1, A2, B1, and B2, with all those in the A1 quadrant being the "meat and potatoes" of the kaizen event activities.

After the countermeasure sticky notes are initially positioned, the team should revisit the assigned locations to ensure that they are indeed in the right quadrant. This is where the sensei will often challenge and push the team. The team will often have some countermeasures inappropriately positioned. This is due to technical and/or behavioral gaps, characterized as follows:

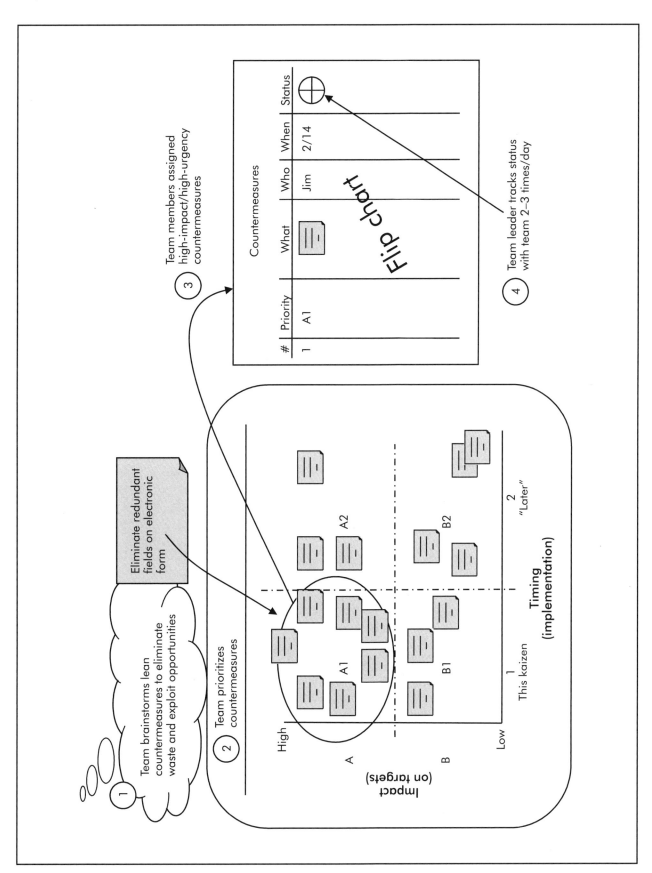

Figure 6-8. Identification, prioritization, assignment, and tracking of countermeasures.

1. Technical oversights can drive improper placement between 1 versus 2 and A versus B due to a misunderstanding in sequence/primacy and/or degree of implementation difficulty. For example, a lack of technical understanding may drive a team to establish a supermarket during the kaizen and delay until after the kaizen the reduction of a painfully long setup time. Experience, however, would tell the team that the setup is so ugly that it must be addressed before the kanban is sized, otherwise the supermarket will be huge and unwieldy. Similarly, a team may not appropriately equate developing standard work with a kaizen target of increasing productivity and thereby mistakenly put the related countermeasure somewhere in the B quadrant.
2. Behavioral issues, such as timidity (maybe laziness) or over-aggressiveness, are usually reflective of improperly selected implementation timing.
3. The combination of technical and behavioral gaps, as no surprise, multiplies the challenges.

Once the team and sensei are reasonably comfortable with the categorization of the sticky notes, they should determine the relative sequencing of the A1 notes, knowing that often the best strategy is to "divide and conquer." This is reflective of the reality that countermeasures often can be implemented in parallel. The next step is for the team leader to assign the A1 notes (and B1 and A2 notes as time permits) to team members, either individually or in sub-teams, along with the day, and sometimes hour, or at least delineation between a.m. and p.m. that the countermeasure is to be completed.

The countermeasure assignment can be visually effected by simply moving the appropriate note to a flip chart countermeasure sheet as in Figure 6-8. If it is desired to display and track the countermeasures in a small space, overhead transparencies or, worst case on a computer, then an 8-1/2 × 11-in. form as in Figure 6-9 may be used. The progress of each assigned countermeasure must be tracked multiple times a day by the team leader to facilitate focus and timely execution. At the conclusion of the kaizen event, the appropriate open countermeasures should be transferred to a kaizen newspaper form.

It is important to note that as the kaizen event progresses, and thus insight into the issues and opportunities at hand increase (in fact, some countermeasure may fail or perform marginally in the "trystorming" phase and need to be jettisoned or retooled), it is probable that new countermeasures will be identified. These countermeasures should be prioritized in the same manner as the initial population and the high priority ones assigned for completion.

Strategy

The initial kaizen strategy is primarily focused on the best approach, within the defined scope, for determining the pre-kaizen situation so that the waste and related opportunities can be identified. With that, the strategy should reasonably anticipate the broad types of countermeasures that may be necessary to achieve the kaizen targets. For example, the process mapping of a complex, long lead-time administrative process will often highlight multiple, non-value-added hand-offs, large variations in process/procedures, and a lack of insight into the status of the "product." This, in turn, will drive countermeasures that are in the realm of standard work, rationalization of hand-offs and thus the reasons for them, visual controls, etc.

True and deep insight into the process is often not obtained until after the pre-kaizen situation is defined through direct observation and the use of the appropriate tools. Often the statement, "We didn't know what we didn't know" can be joined with "... when we were pre-planning and kicking-off the event." So, based upon the newly obtained knowledge, the kaizen scope and targets may have to change. For example, the scope may be way too big and/or the ability to determine the root causes of the waste may require the deployment of some rather time-consuming tools that will require much data collection, analysis and testing. In such a situation, the team may have to reduce its scope, targeting possibly fewer processes or even a sub-process. Sometimes the targets may be revised to only reflect the establishment

Team name: FNOL				Date: 6/22/YY			
	Countermeasures			Page: 2 of 2			
	Observation/Waste Identification Opportunity	Action to be Taken	Person Responsible	Due Date	Percent Complete			
11	Representative opens system screens and then backtracks 3+ times during first notice of loss (FNOL) from insured	Develop standard work, best, least-waste screen sequence for FNOL process	Timlin	6/23	25	50	75	100
12	Not evident when a representative (with headset) is on a call and, if so, whether it is with a customer	Implement simple, low-tech, worker-managed visual control to indicate phone status	Fontana	6/22	25	50	75	100
						50	75	100
					25	50	75	100
					25	50	75	100

Figure 6-9. Countermeasures form. (For a blank form, see Appendix A.)

of standard work (because the process is long and complex and there is nothing that resembles standard work in the pre-kaizen situation).

The scope and strategy dynamic is reflected in Figure 6-10. The level of waste in a target area, which many times is driven by the extent of its lean penetration, dictates the breadth and depth of the kaizen. The hierarchy of needs, at the risk of gross oversimplification, is typically:

1. basic process stability and availability,
2. flow, and
3. pull.

Revising a kaizen strategy does not indicate failure. It is just common sense. However, when such a situation is encountered, it should be reviewed with the sensei and discussed and approved at the daily team leader's meeting or sooner.

Improvement Ideas

Kaizen events facilitate the flow of ideas. Many ideas, for a specific problem or opportunity, ultimately boil down to the best single "implementable" idea. From the perspective of the storyline, improvement ideas represent the big and medium sized "wins" achieved by executing key countermeasures. Accordingly, this storyline element serves as a bridge between the countermeasures and the post-kaizen situation. The improvement idea storyline element also functions as a vehicle for simply communicating the *implemented* ideas (in the worst case, where long lead-time items preclude finalization during the kaizen event, the idea may be simulated/trialed, to prove-out the concept and impact).

In typical lean fashion, standard work for communicating an improvement idea is effected through a standard form. This form is aptly named the "improvement idea form" as shown in Figure 6-11. Sometimes it is called a "kaizen reporting sheet." The form, a one-page document, is designed with two purposes in mind, it: 1) promotes and reinforces scientific thinking, and 2) summarizes the substantive idea so it is quick and easy to document and understand.

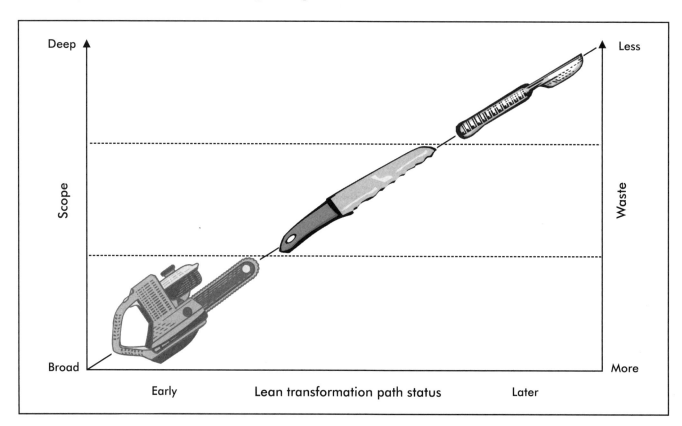

Figure 6-10. Kaizen scope, waste, and approach.

Improvement Idea

Kaizen scope: Outpatient blood draw   Team name/#: GH-2-5/YY   Completed by: Alfred Krueger   Date: 5/25/20YY

Description of Problem	Description of Steps Taken	Results
Walk-in patients wait an average of 20.6 minutes throughout the blood draw process.	Combined registration and blood draw operation to achieve continuous flow. To support this improvement: 1) improved work area design, 2) developed standard work, and 3) created a cross-training plan (several phlebotomists need to learn the registration process).	Reduced patient wait time by 68%. Reduced work content by 9.5%.
Double and triple "handling" of the patient from point of (volunteer) greeter to registration process to actual blood draw.		
Discontinuous flow and no work balance (% load) between registration and blood draw.	Implemented a FIFO lane with related visual controls between greet and registration/blood draw.	

Before Kaizen

Blood draw—the waiting room is for "patience"

Greet, copy license and insurance card → 35 sec. → 85 sec. → Registration → 720 sec. → 120 sec. → Blood draw → 480 sec. → 300 sec.

Average queue/wait time = 20.6 minutes
Process time = 8.4 minutes

After Kaizen

Greet, copy license and insurance card → 35 sec. → 75 sec. → Maximum: 4 —FIFO→ Registration and blood draw → 360 sec. → 380 sec.

Average queue/wait time = 6.6 minutes
Process time = 7.6 minutes

Figure 6-11. Example improvement idea form. (For a blank form, see Appendix A.)

> ### "Trystorming"—Spinning the PDCA Wheel
>
> Most people are familiar with brainstorming. It is typically used in the context of a somewhat organized process by which a group of people share thoughts or ideas, often to solve a specific problem. By design, the process yields diverse, occasionally "off the wall" ideas. But usually there is a diamond among the many; sometimes it is a great singular idea or several ideas or a hybrid that can be cobbled together from multiple ideas. The problem is that ideas represent only a "plan." Plans cannot be translated into action, validated, and improved without the rest of the Deming wheel—do, check, and act. Enter "trystorming."
>
> "Trystorming" represents a melding of both brainstorming and simulation. Simulation and its induced learning and adjusting provide the "try" within "trystorming." The production preparation process (3P) makes heavy use of "trystorming." While 3P "trystorming" is more formal than that used in the typical kaizen, it is worth reviewing.
>
> The use of 3P prescribes identifying seven alternatives or seven different ways for product or process design. While there may not be any special magic in the number seven, it does pull a multitude of ideas from a team. Team members are encouraged (individually) to think outside of the box. They are asked to use child-like simplicity to highlight key words that represent what they are trying to do within the process or product design and to borrow ideas from nature (for example, how a turtle's shell *encapsulates* its body) to represent those key words. Along with capturing these key words or concepts, they are required to sketch them and with this, understand their operation.
>
> The team filters the individual member ideas down to just seven. Using simple, crude materials and methods—cardboard, duct tape, wood, foam, etc., the team begins to "try" the seven ideas through simulation. While the simulations are not expected to be perfect, they should be critically observed and measured. From the insight garnered from this experience, the team will better understand how to improve the process or product design. The three best design concepts as measured against the pre-established design criteria (cost, quality [including dimensions and performance], delivery, etc.) are then chosen. These three designs are subjected to more rigorous "trystorming" activities in which problems will be fixed on the spot and learning and insight will continue to grow. It is from this experience that the best design will be selected.
>
> Virtually every kaizen should employ a level of "MacGyveresque trystorming." It is the only tried and true way to quickly engage the team and generate, test, and improve on a number of different ideas, ultimately validating the one best way within the finite span of a kaizen event.

This is not quantum physics, but nevertheless the improvement idea form does prompt a level of scientific thinking. It requires a description of the problem, which infers the identification and understanding of the root causes, and the steps or actions taken (countermeasures). Both of these elements are derived from the storyline's pre-kaizen situation and countermeasures. The form also requires the kaizen team to quantify the results or impact of the steps taken. These results must be validated, if at all humanly possible, through direct observation.

In the bottom half of the form, there are two sections requiring the characterization of the before and after situations. Think of these as the idea-specific pre-kaizen and post-kaizen situations or conditions. The form's five sections are:

1. problem,
2. steps,
3. results,
4. before condition, and
5. after condition.

These are virtually identical to those within the A3 report. The only significant difference between the A3 and the improvement ideas form is in the application. The A3 is an active planning and execution tool, while the improvement idea form is more historical in nature.

The simplicity of the improvement form's one-page, five-section design forces the author to be concise and, at the same time, enables the audience to digest the message at a glance. There is clearly a balance of substance and simplicity. The use of cartoon-like drawings or sketches that often represent the before- and after-kaizen situations does not mean that the team is not taking things seriously. It is quite the opposite.

In the heat of the kaizen battle, no one should be spending a great deal of time preparing or reviewing elegant charts, graphs, and slides. Instead, the team should be focused on eliminating waste. Similarly, during the team leaders' meeting where the improvement idea form should be a standard way of communicating what has been accomplished in the last 24 hours, the participants should be able to quickly understand the message. In this way, management, the kaizen promotion officer, and the sensei can more efficiently get to the job of coaching and supporting the teams and let the team leaders get back to leading their teams.

Post-kaizen Situation

The post-kaizen situation is documented much like the pre-kaizen situation. The difference between the two storyline elements, however, is two-fold in nature: 1) the new situation should be much "leaner" than the old, and 2) the missions are totally different.

The improvement over the prior state to the now, hopefully realized future state is the result of successfully executing high-leverage countermeasures. The mission of the pre-kaizen situation was to identify waste and the related opportunities—essentially a PDCA focus on improvement. However, the post-kaizen situation is primarily about standardize-do-check-act (SDCA). In other words, without standardization, the hard-fought gains cannot be sustained. Standardization is incomplete without:

1. validation of the new standard work,
2. training individuals on that standard work, and
3. implementation of a lean management system (LMS) or the expansion of a pre-existing LMS.

Another word for "situation" in lean speak is "reality." To ascertain the reality of something, direct observation is necessary . . . over and over again. This means that the post-kaizen situation, prior to the report-out, must be validated, the

Gemba Tales

Try + fail + learn = improve. While working with an inexperienced kaizen team, it was clear early on as to why there was little traction. Each observation and improvement idea went through the "bobble head" committee. The team was not empowered to act without consensus from the engineering manager, production manager, and the environmental, health, and safety manager. The committee ostensibly was looking for guaranteed results before giving the green light to implement any improvements. This paralyzing management method mirrored an organization that suffered long delays, a slow rate of change, limited results, and limited sustainability. Something had to change! The kaizen team needed to get some quick wins to gain some ground on the productivity targets.

On the second day of the kaizen, we decided to pull the engineering manager into the team and encouraged her to spend some quality time on the shop floor with us. We spent the rest of the day at the gemba with the operators and team, conducting direct observation, discussing ideas and alternatives, creating a frenzy of "trystorming" that extended late into the evening. By the end of day two, we had developed two prototype tools and three standardized process improvements that generated a 40% productivity improvement. The engineering manager was so excited about the quick turnaround and improvements that she cleared her calendar for the rest of the week to help the team and get her hands dirty again. It seemed that no idea was too far fetched.

By the end of the kaizen, the team had established an 85% productivity improvement with virtually no action items to complete. During the event report-out, the engineering manager shared that the experience of implementing real-time solutions with excited and engaged operators rekindled the enthusiasm that she had as a junior engineer years previously.

Tale shared by Edward P. Beran

unforeseen roadblocks addressed (for example, operator balance issues, previously unobserved process variation, information and material flow problems, etc.), the frontline supervisors trained, and the operation must run the new standard work for at least portions of all shifts.

The absolute criticality of validating the post-kaizen situation requires a type of back-scheduling (and sometimes a limitation in event scope). For example, if the report-out is midday Friday, then the new process must be running by midday Thursday. To extend this example, this may require managing the team so that while most of the team members are wrapping up countermeasures and other tasks early in the afternoon on Thursday, a few team members are observing, documenting, and training frontline supervisors and area leaders, and fixing problems as they arise. This observation and training strategy should continue with the second and third shift as required. Friday morning the team should review the plan versus actual performance and identify and implement any necessary countermeasures.

Performance vs. Kaizen Targets

The performance versus kaizen targets storyline element is straightforward. It is purely for the purpose of reporting how the team performed against the initial kaizen targets or, in the situation where management has reached consensus on a mid-kaizen shift in targets, the revised targets. While this may seem like a perfunctory, "Let's wait until the end of the event to see how we did" thing, it should not be. When properly incorporated into the team leader meeting process and work strategy, this storyline element is much more dynamic.

Performance versus the kaizen targets should be reviewed on at least a daily basis. This is the reason why the kaizen event target sheet (Figure 5-12) reflects columns for each day of the event. It is also consistent with the lean notion of small and thus frequent management time frames (think of concepts like takt time, pitch, and plan versus actual reporting) in which abnormal conditions can be identified almost real time and countermeasures rapidly deployed.

As discussed in Chapter 5, under certain conditions it is more appropriate to develop and track to a kaizen mission statement. Performance versus the mission statement can be reflected by means of checkmarks or colors (green or red) to indicate whether targeted future-state characteristics have been achieved. As within the target sheets, quantitative mission statement elements should be compared to actual post-kaizen measured performance.

Kaizen Newspaper

When people hear the words "kaizen" and "newspaper" together for the first time, there is often a ponderous look on their face. They may grasp "kaizen" and undoubtedly understand that a newspaper is a daily or weekly publication that contains current news. But what is a kaizen newspaper?

The kaizen newspaper (see Figure 6-12), also called a kaizen follow-up list, 30-day list or 40-day list, is a means for identifying, assigning, and tracking those critical countermeasures that remain unfinished at the end of the event. The kaizen newspaper ostensibly gets its name from the fact that to facilitate timely execution of assigned countermeasures it, like a regular newspaper, should be "read" on a daily and, certainly no less than weekly, basis. Fortunately, however, the news within this newspaper is meaningful and actionable and thankfully there is no place for editorials, feature articles, and advertising. Note that flow kaizen events do not use a kaizen

> **Pro Forma Results**
>
> In some situations, when the process cycle times are long (hours or days), the lead times are long (days, weeks, or months), and/or there is limited customer demand, it is difficult to validate the new improvements and the related standard work. While much can be inferred from simulation, whether with real parts and real machines, or desktop "walk-throughs" of the standard work with old files, transactions, contracts or claims, etc., it is not the real thing. Only time will truly tell. Accordingly, the kaizen event results are what may be called *pro forma*, essentially "as if" the improvements work as anticipated. For this reason, conservativeness in estimating the performance is warranted. Additionally, such results should carry a big asterisk with an appropriate footnote, explaining the situation. The post-kaizen audit should seek to verify the results within 30 days after the kaizen.

Kaizen Newspaper

Team name: Unscheduled maintenance service
Date: 9/14/YY
Page 1 of 1

	Problem	Countermeasure	Person Responsible	Due Date	Percent Complete			
1	No standard means for determining service work content or duration by craft • Makes capacity analysis and establishing plan vs. actual extremely difficult	Develop parametric work content "calculator" to be applied to type and quantity of unscheduled maintenance event	Kaplan	9/28	25	50	75	100
2	No visual controls for designated location of kitted and pre-staged service parts	• Install visual controls at service bays 2 and 3 reflecting specific location of kitted and pre-staged service parts • Include means for clearly indicating the assigned unit for which the parts are picked, maximum number of pre-staged jobs, status, etc.	Shewhart	9/21	25	50	75	100
3					25	50	75	100
4					25	50	75	100
5					25	50	75	100

Figure 6-12. Kaizen newspaper. (For a blank form, see Appendix A.)

newspaper. Instead, the detailed follow-up action items are captured within the value stream improvement plan.

The kaizen newspaper must reflect all incomplete A1 (high impact/quick implementation) countermeasures. Clearly, final posting and implementation of standard work fits within the category of A1, even if it has not been formally flagged as such. Make no mistake, long kaizen newspapers are an indication of a kaizen that was either scoped incorrectly, staffed by an ineffective team that did not get things done, or both.

Depending upon their continued relevance (judgments do change throughout the course of the kaizen) and required investment and lead time, A2 (high impact/longer implementation) countermeasures also should be reflected in the kaizen newspaper. Each A2 should be assessed, after estimating the benefit of its implementation and the related cost, with management to determine whether and how it should be pursued.

As inferred by the two other names for the newspaper, 30-day list and 40-day list, newspaper item completion is time bounded. The list should be of things that can and should be reasonably completed within 4 to 5 weeks after the kaizen event, if not much sooner.

The effective team leader will exercise care to ensure that the newspaper language is precise and understandable and that the persons assigned are responsible and at least committed to their newfound ownership—both task and timing. Countermeasure owners should be able to "negotiate" a due date with the team leader that is reflective of level of effort, authority, and the need to execute the countermeasure promptly.

For example, the description of the problem may be, "Associates must physically seek out team leader for *all* credit approvals." The corresponding countermeasure could be, "Analyze credit request/approval history, formalize criteria and standard work, and empower team members for low-risk decisions." This countermeasure can only be executed with direct input from the workers. So, it makes sense to assign the countermeasure to the team leader or manager who participated in the event, with the understanding that she will engage and include the associates who work in the department or cell, especially those who participated in the kaizen. Only one name should appear in the "person responsible" column to facilitate accountability and easy status review. A countermeasure of this type should be completed within a week, so the due date would be one week from the date of the assignment.

The only exception to the 30- to 40 day newspaper execution horizon is when the countermeasures are more deliberate and complex, requiring multiple time-phased interdependent steps, possibly including a pilot phase and beyond. These situations include, for example: 1) prove-out of pro forma standard work, possibly in a production preparation process (3P) type application, where there may be long lead-time items for post-prototype equipment or tooling, or 2) the rollout of a new standardized, least-waste way of tracking performance measurements, which will be piloted for 4 to 6 weeks, then deployed to different locations, functions, and levels. The level of planning in these types of situations often requires a project plan/Gantt chart reflecting steps, timing, ownership, and level of effort as well as key milestones and checkpoints to facilitate risk management (schedule, cost, and technical) and, ultimately, execution. The kaizen newspaper should reference the project plan or in many instances the project plan *is* the newspaper.

For obvious reasons, the newspaper is one of the last things completed before the report-out. The risk, therefore, is that it becomes a perfunctory exercise, a small speed bump between wrapping up the post-kaizen situation and preparing for and conducting the final presentation. Because of this dynamic, the newspaper should be subject to an immediate post report-out review (Chapter 7 discusses this further).

Lessons Learned

Just prior to the report-out preparation, which should be no more than several hours before the time of the presentation, each team member should briefly reflect upon the kaizen event. This is a rather abrupt change in gears as the team has been charging, head down, toward the finish line. Yet, this reflection, not inconsistent with the Japanese notion of hansei, should be conducted

> **Socrates would be Proud**
>
> *Hansei*, a Japanese term meaning reflection to acknowledge mistakes and pledge improvement, extends to both the personal and corporate or team level. In this spirit, after key milestones are reached, it is beneficial to reflect on past performance and identify the shortcomings of the project, initiative, or event and develop countermeasures to avoid repeating the same mistakes.

after an important milestone, namely the completion of the execution phase of the kaizen.

The reflection is intended to identify, articulate, and share as a team what was "learned" and, if need be, what to change in the future to better the process. The lessons are often multi-dimensional and encompass individual and team discoveries relative to the kaizen process, lean, team dynamics, change management, the targeted process (kaizen scope), personal growth, etc. This is an opportunity to affirm as individuals and as a team what has been experienced. A broader audience, via the presentation (yes, it is one of the agenda items), is also given insight into what the kaizen experience is about.

The process to identify the lessons learned is rather simple. After the team is provided with a brief explanation of the process and each person has been given several moments to gather his thoughts, the facilitator, flip chart at the ready, asks for someone to name a lesson learned. This lesson learned is recorded, essentially word for word, unless clarification is needed. In fact, there should be no questions (or arguments) from teammates other than for clarification purposes.

The facilitator proceeds, team member by team member, around the table recording each lesson learned. If a participant does not have one at the ready, he can simply say "pass." Once several rounds have been made with the only response being "pass," the sharing is completed. Typically, team members are not at a loss for words.

Occasionally, a team member will raise issues or opportunities for improvement. For example, "Pre-planning could have been more effective—next time identify and secure target claim files for possible process mapping prior to the event." The kaizen promotion officer, in such a situation, should take note of these and incorporate this notion in future pre-planning.

TEAM LEADER MEETINGS

Team leader meetings are conducted mid or late afternoon of each day of the kaizen. The exception is the final day when the report-out is made. Often the last team leader meeting of the week is held at the gemba to review and demonstrate the improvement activities that have been accomplished throughout the kaizen. This can be conducted in an "open-house" format, within which management and members from other kaizen teams can "drop by" during a prescribed time period to witness things like newly implemented standard work, simulations, visual controls, etc., and ask questions.

The traditional (non-open-house) team leader meeting is comprised of a 10- to 15-minute interactive presentation made by each team leader and co-leader to management, the sensei, and

> **Example Lessons Learned**
>
> - "Data tells the story. Different from perception."
> - "Must take time to go through the process to get the data."
> - "Can't do this part-time (for example, meetings twice a week)."
> - "Amazed at how much of the lean technology I had forgotten."
> - "Amazed at the effectiveness and creativity of the team."
> - "Reinforces need for intense and disciplined attack of problems and opportunities."
> - "Provided a forum for decision and agreements to be made in a set time frame . . . and then acted on the decisions."
> - "Process broke down hesitancy . . . just did it. Necessity is the mother of invention."
> - "Benefits are indisputable."
> - "We can take this process to other areas."
> - "Did not overanalyze things. Got stuff done."
>
> *Actual quotes*

> **Keep the Meeting to Takt Time**
>
> The discipline of keeping the meeting to takt time respects the time of the management attendees and gives them one less excuse not to be there. It also preserves precious time for the team leaders so they can get back to their teams and keep making improvements.

any out-of-town guests. Its purpose, at a broader level, is to:

- Reinforce the rigor of the scientific kaizen process, which is implicit in the strategies and forms (pre-event area profiles, target sheets, countermeasure forms, improvement idea forms, etc.) that aid the identification, acknowledgement, and elimination of waste.
- Hold the team leader, and therefore the team, accountable. Pressure, properly applied, is beneficial.
- Engage managers and make them do their jobs as lean leaders.
- Facilitate the removal of barriers.
- Ensure team and leadership alignment with the kaizen targets.
- Facilitate and develop better lean thinking.
- Avoid surprises.
- Drive results.

At a more tactical level, the purpose of the team leader meeting is to quantitatively and qualitatively communicate targets, strategy, activities, accomplishments, barriers, and countermeasures, while also providing a forum for feedback and suggestions, praise, and occasionally chastisement from the "audience." The presented items are typically and necessarily daily in nature—a 24-hour retrospective when talking about activities and accomplishments and a 24-hour plan of future activities that must, by definition, have the end in mind.

The agenda is reflected in Figure 6-13. The meeting can be conducted in one of two ways: 1) in a designated room, distinct from the team break-out rooms and sized to accommodate the presenters and the audience, outfitted with an overhead projector for "slides" copied on acetates (use of an LCD projector should be resisted as it encourages the waste of over-processing—the creation of "pretty" slides), and sufficient wall and table space to position any relevant flip charts, props (for example, a prototype fixture, new form, etc.), and digital photos, or 2) with the audience and presenters meeting in each team's break-out room and the team leader referencing the appropriate forms, flip charts, photos, props, etc., already hanging in storyboard progression on the break-out room wall. If option 2 is selected, then the team members should not be in the break-out room during the team leader meeting. Instead, the team members should be executing countermeasures, conducting observations, etc., or on an appropriately timed break.

In essence, the team leader meeting is a formal PDCA forum in which management and the sensei can serve as a checking function relative to team strategy, progress, identification of barriers (real or otherwise), opportunities, and the scientific manner by which the team is letting the data lead them. It is within this context that management and the sensei can assist the

> **Should the Team Co-leader Attend the Team Leader Meeting?**
>
> There are two schools of thought regarding co-leader attendance at the team leader meeting. Often people exclude the co-leader from the meeting so that he can remain with the team members to direct and manage them. However, if the team's daily plan and related assignments are explicit and it has one or more grizzled kaizen veterans, this may be an opportunity to expose an inexperienced co-leader to the team leader meeting process. Team leader meeting participation can serve as a training and development experience for the co-leader, thus helping him prepare for a future role as a team leader. Further, participation can help "calibrate" the co-leader by providing firsthand experience of management's focus and sense of urgency, something that he can take back to the team. As is often the case, the co-leader may be experienced, having recently served as a team leader of one or more kaizens. In such a situation, he can assist the leader with meeting preparation and delivery.

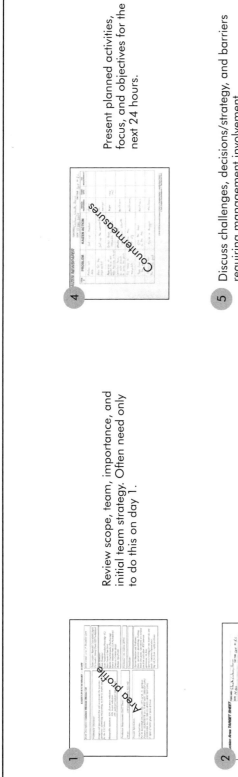
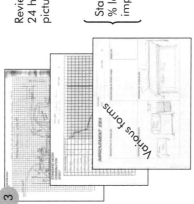

Figure 6-13. Team leader meeting agenda (15 minutes).

team leader in the act/adjust phase, by means of suggestions to:

- hold or alter the current team strategy and/or tactics and possibly the current kaizen scope,
- introduce new tools and approaches,
- challenge current thinking as to what is acceptable and what is possible, and
- facilitate the removal of certain barriers (for example, by means of a not too subtle paradigm-altering discussion with a certain non-lean thinking upstream operation, assigning needed maintenance resources to the kaizen, etc.).

Essentially, the team leader meeting is an opportunity for management to serve as a team leader to the team leader. The team leader is called to provide the team with the requisite encouragement, freedom, and resources, while challenging the team members and holding them accountable to successfully achieve the clearly articulated goals. Implicit in the team leader meeting is the passionate reinforcement of a sense of urgency and commitment (why else would management spend time each day in these meetings?). Additionally, the meeting process teaches the team leaders valuable presentation preparation and delivery skills and enables management to interface with and encourage team leaders, and informally evaluate their skill sets. Last, but not least, the meeting serves as a means of communicating kaizen progress and changes in a near real-time manner, which will help avoid final report-out bombshells on the last day of the kaizen that may be harbingers of unsustainable improvements. For example, the process should help avoid a situation where a key manager maintains that his area has been "blindsided" by the new changes.

WORK STRATEGY AND TEAM EFFECTIVENESS

Work strategy drives team effectiveness and, ultimately, results. As reflected in Figure 6-14, it operates within the intersection of the kaizen storyline and the team leader meeting process, while drawing upon the core and technical competencies of the team and its members. The foundation of work strategy is pre-event planning. Not surprisingly, work strategy success is largely secured by the leadership skills of the team leader and is highly influenced by the sensei.

Any shortfall in team effectiveness is exacerbated and magnified by the extremely short duration of the kaizen event. Team effectiveness is a measure of how the team (of individuals) works together during the kaizen. From a quantitative perspective, the team is expected to meet or exceed the kaizen targets. From a qualitative perspective, it is hoped that the team does this in an effective and efficient manner, truly leveraging the talents and experiences of the various team members and achieving something greater than what they could have as just a loose confederation of individuals.

The employees are the engine of a successful lean transformation. Employees accomplish more and are more fulfilled when participating and contributing as members of effective teams. Team skill development benefits are therefore not limited to only kaizen, but transcend it and extend to daily operations and the very culture of the company.

> **Gemba Tales**
>
> **Forum for proactively confronting and dealing with reality.** The kaizen promotion officer told the sensei that there was no sense in having team leader meetings because most of the management team would not attend. Aside from corporate support, there was little here at this particular location. *Red flag!* By day three, the change management issues that the kaizen event was generating were not trivial—the redeployment of two of the three operators and the abrupt change from a manufacturing resource planning (MRP) driven schedule to a mixed model kanban system. And, oh yeah, one of the people being redeployed was the wife of the planning supervisor (the MRP guy)! What occurred in fairly short order was a *defacto* team leader meeting to hammer out these things. Moral (at least one of them) of the story: team leader meetings are non-negotiable for a reason.

The foundation of team effectiveness is solid pre-event planning. It all starts with event selection and then moves to the all-critical team member selection based upon a mix of experience, perspective, core and technical competencies, value stream and process representation, and change management objectives. Thoughtful team selection must be supplemented by things like good communication, pre-work and logistics, etc. In other words, teams must be set up to "win"—assigned reasonably scoped events, possessing the right information, the proper supplies and equipment, and receiving timely and relevant training on lean fundamentals and specific lean tools and techniques. However, these things alone are no guarantee of team effectiveness. The team must, among other things, employ team effectiveness promoting behaviors.

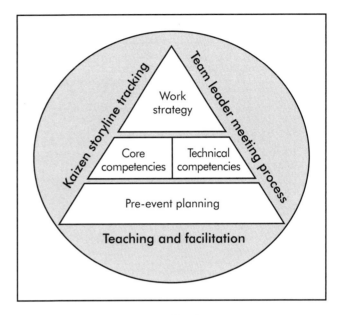

Figure 6-14. Kaizen team effectiveness.

Core and Technical Competencies

Skills exist on two planes. The first, and the one which people tend to focus on primarily, is technical skill. The second, often overlooked skill set is core, often behavioral in nature. Kaizen team effectiveness demands both types of skill.

Technical skills encompass, in the realm of kaizen, how proficient a person is in areas like observing reality and documenting it on the various standard work forms, or sizing kanbans, or possibly identifying and developing autonomous maintenance activities, etc. These are all teachable skills and improve with proper coaching and experience. Behavioral skills are similar, but because of their "background" status in people's psyche, they are best facilitated by effective team leader behaviors. The team leader must articulate and model the proper team member behaviors and periodically audit how the team is performing. From this, the team leader, in conjunction with the team, can develop countermeasures to close the performance gaps.

There are several things that the team leader must provide to create an environment in which the team can be effective. These "inputs," reflected in Figure 6-15, are intimately linked with the team ground rules and behaviors. The desired team environment is shaped by work strategy techniques, reinforced by the team leader meeting process, and guided by the storyline.

The team leader inputs are supplemented by certain basic kaizen event ground rules or agreements. These rules, listed in Figure 6-16, are a little more earthy than the 10 + 1 principles of kaizen reflected in Chapter 2 . . . although MacGyver is alive and well in both. The rules lay the groundwork for effective teamwork. At the beginning of each event, the team leader should emphasize the ground rules. It is helpful to post them in the kaizen break-out room so that they can be easily referred to and reinforced as needed.

While the ground rules are critical, there are other process-oriented behaviors (or behavioral skills) that are hallmarks of an effective team. These behaviors, reflected in Figure 6-17, help provide focus and harness the skills and energy of the team. It is useful to review these behaviors with the team at the outset of the kaizen and then periodically conduct a process check. The process check, which is best facilitated by a kaizen promotion officer or sensei, generates an assessment or performance ranking for each behavior, by individual, in a roundtable voting process.

The process check results will reflect: 1) team consensus (tight distribution) somewhere along the scoring continuum (high, medium, or low),

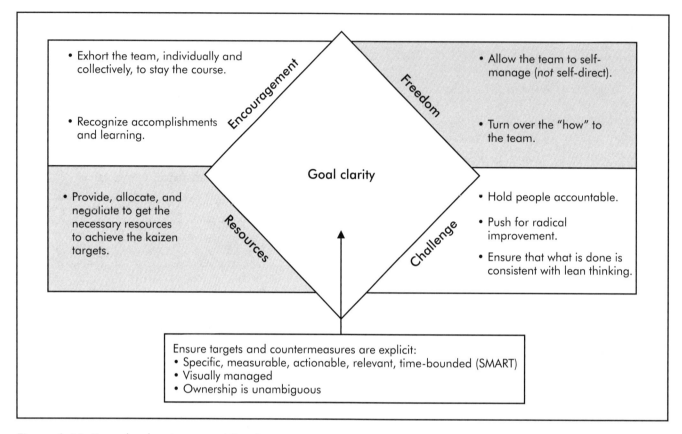

Figure 6-15. Team leader: team-enabling inputs.

or 2) diverse scoring, which, more or less, reflects differing opinions on how the team is performing in a specific area. The facilitator should probe low or medium consensus scores and help the team articulate/identify the root causes and the countermeasures/change in behaviors required to improve performance. Also, the facilitator should probe the scoring of "outliers" by inviting these individuals who are on the high or low end to explain their perspectives. Again, the team should identify the root causes and the related countermeasures. To keep things moving along, it is suggested that the team work on no more than two to three gaps at a time.

Sense of Urgency—MacGyver Meets the Fruit Fly

If there is one behavior that must under-gird every kaizen event, it is a sense of urgency. It is reflective of the "Just *do* it . . . now!" kaizen ground rule. How else can big improvements be made within a short time span of 3 to 5 days? This notion is often lost on first-time kaizen participants because their frame of reference is the current state—the current business processes, current performance levels, and current pace of

---

Kaizen Ground Rules

- Leave all titles and ranks at the door.
- Treat others as you would like to be treated.
- Improvement requires *change*. Don't waste time justifying the current situation.
- Keep an open mind.
- Maintain a positive attitude.
- Deal from data, *not* perception or emotion.
- Create a blameless environment.
- There is no substitute for hard work
- Plans are useful *only* if they can be implemented and if the gains are sustainable.
- Just do it . . . now!

Figure 6-16. Kaizen ground rules.

1. Rigor	1	2	3	4	5	6	7	8	9	10
	No discipline, not on track							Following kaizen standard work and using effective work strategies		
2. Urgency	1	2	3	4	5	6	7	8	9	10
	Limited or inconsistent effort to get the right things done . . . no fire							Everyone is doing whatever it takes to satisfy kaizen targets		
3. Participation	1	2	3	4	5	6	7	8	9	10
	Few members dominate, some do not participate							Everyone contributes and is involved		
4. Listening	1	2	3	4	5	6	7	8	9	10
	>1 Person talks at a time, repetitions, interruptions, and side conversations							One person talks at a time; clarifying and building of ideas		
5. Leadership	1	2	3	4	5	6	7	8	9	10
	No attempts to bring team back on track and encourage equal participation							Team members intervene to keep team on track and manage equal participation		
6. Decision quality	1	2	3	4	5	6	7	8	9	10
	Team decisions based on opinion and inferior to individual assessments							Team expertise and decisions are data-driven and superior to individual judgments		
7. Candor	1	2	3	4	5	6	7	8	9	10
	Team members do not freely share thoughts							Team members speak openly		
8. Fun	1	2	3	4	5	6	7	8	9	10
	Rather have a root canal							☺		

Figure 6-17. Team behavioral audit. (Also see Appendix A.)

change. This is where the marathoner meets the sprinter . . . in a 100-yard (91-m) dash.

The concept of urgency must be constantly reinforced. As the company's lean transformation accelerates and demonstrates staying power, the culture will begin to embody a sense of urgency. But, early on this will not be the case. Accordingly, kaizen team members must hear, and see, a consistent message of urgency. This message must be "broadcast" from their time of selection for the team, to the kick-off meeting, to pre-event training, and every hour of every day during the event. This message must come from senior management, kaizen team leaders, co-leaders, the kaizen promotion officer, the sensei, and fellow teammates. Further, it must be ingrained within the work strategy.

The lifespan of the fruit fly can give people a perspective of urgency (see Figure 6-18). The fruit fly lives for about 2 short weeks. With this in mind, the team can envision the first week requiring incredible intense focus and work

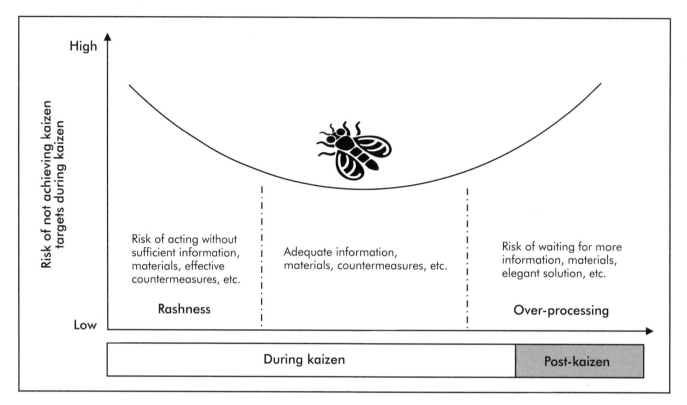

Figure 6-18. Urgency balance.

(the kaizen event) and then the second week providing an equally long retirement. This new paradigm suddenly changes the way lead times are looked at and how things get done. For example, the previously "acceptable" 7-day lead time for developing a new visual control for tracking transaction status suddenly seems totally outrageous. "Can't we get it done quicker? A prototype needs to be in place in 24 hours! Does the solution have to be so stinking elegant? Can't we make something using easily accessible, cheap stuff? Better yet, can we salvage something from somewhere else in the facility?" (Kaizen shopping = beg, borrow, steal, or rent before buying.) This is where MacGyver meets the fruit fly. The 2-week lifespan and 5-day kaizen duration means the team must be incredibly creative. Of course, there is tension between acting without sufficient information and waiting/working to get still more information before acting. Similarly, there is tension between purchasing something (modest) for the kaizen (for example, tooling or supplies) and trying to maintain the usual short-sighted fiscal restraint. This is where simple cost-benefit analysis must be used. (And, spend the extra $25 to get the thing express delivered so that it can be used or installed during the kaizen!)

Gen. Douglas MacArthur submitted that every military defeat could be summarized into two words, "too late." The combination of the proper sense of urgency, experience (that of the team, kaizen promotion officer, sensei, team leader, and meeting participants), adherence to the storyline, and a well-executed work strategy will enable good decisions and avoid the trap of being too late.

Work Strategy

Work strategy is often where teams "lose their way." It drives focus, intensity, and execution, including the what, when, how, and by whom things are worked on. Merely knowing how to conduct excellent observations, fill out the forms, and perform some great analyses does not guarantee results.

>  **Gemba Tales**
>
> **MacGyver would have been proud.** A kaizen team was tasked with improving the throughput of a cooler. The cooler, an insulated, liquid-cooled vessel with a rotating blade, was constraining the entire line. The upstream process was feeding a compounded material that had a high temperature because of the friction generated during the previous mixing operation. After some direct observation, it became evident that the cooler could not cool the material within the takt time. On top of that, there was a lot of moisture within the vessel that further complicated the process and made the vessel troublesome to clean.
>
> The team distilled the hypothesized root cause, with the aid of cause-and-effect diagramming, down to temperature and moisture. Question: What reduces temperature and moisture? The answer: An air conditioner. Within hours, the team had secured a roof-top air conditioner from the "bone yard" behind the facility, purchased approximately 100 ft (30 m) of duct tubing, obtained the necessary electrical safety clearances so they could tap right into a nearby electrical box, ran the tubing up to the second floor of the mezzanine where the cooler was located and *voilà*, problem more than fixed. Within several days, the team had proven out a $1 million (annual) solution. The permanent fix came shortly thereafter with the installation of a properly sized and leased air-conditioning unit.

Work strategy, no surprise, is founded upon the PDCA cycle and reflects six basic steps or activities:

1. *Plan*—the initial strategy.
2. Plan/prioritize.
3. *Do*—assign/execute.
4. *Check*—track status and impact.
5. *Act*/adjust—refresh strategy.
6. See—document/visually manage.

These, along with the behaviors and environment previously discussed, help facilitate the "flow" of the work.

There are five optimal worker experiences (Csikszentmihalyi 1990):

1. the worker sees the whole,
2. the worker has a high degree of control and involvement,
3. the task requires full attention,
4. there are no or few interruptions, and
5. there is immediate feedback.

These experiences, in many ways, correlate to what team members should "enjoy" during a kaizen event. A good work strategy helps facilitate this experience.

As reflected in Figure 6-19 and true to the PDCA nature, the six work strategy activities operate in a cycle or progression. This cycle, once launched by the initial strategy, is repeated continuously throughout the kaizen event as the team identifies waste and opportunities with each observation, gains new insight from "trystorming" activities (whether successful or unsuccessful), assesses gaps between current performance levels and the kaizen targets, gets new information relative to closing lead times (for example, the newly ordered hardware has been delayed for 24 hours), responds to new tactics or revisions to the strategy recommended within the team leader meeting, etc. As the team is guided through the event by the storyline, there is a constant "refresh."

Initial Strategy Considerations

There are two distinct activities within the work strategy cycle: 1) implementing the initial strategy, and 2) refreshing that strategy. The first activity is essentially an input to the cycle. This input is representative of the initial strategy that was (hopefully) formulated before the event and communicated during the kick-off meeting. It details the team's approach, "out of the blocks," for the observation of reality and determination of the pre-kaizen situation. This initial strategy and its evolution are critical team leader meeting elements.

Strategies often must be modified. This is typically due to:

- poor pre-event planning,
- new data and insight, and/or
- "scope creep."

One relatively common pre-event planning error is scoping an event that is too large (and sometimes

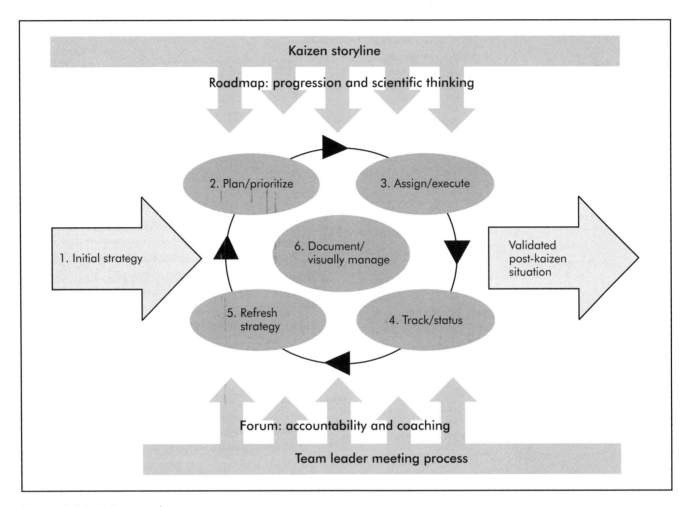

Figure 6-19. Kaizen work strategy.

too small). For example, a single team may be tasked (usually by a delusional overachiever) with developing standard work for a 12-person assembly process with long cycle times and the target of a 40% productivity improvement, inclusive of work area design; establishing a material replenishment system for 9 purchased and 11 internally supplied items; setup reduction of 50% on two machines, etc. In such a situation, the kaizen team is faced with:

- addressing all items in a superficial manner with limited impact,
- reducing the scope and focusing on meaningful and sustainable improvements (for example, just focusing on setup reduction), or
- garnering more kaizen resources from management and establishing an additional team or two to address appropriate right-sized portions of the scope.

This situation is contrasted with mid-event course corrections precipitated by the collection of new data and insights. For example, it may be determined that 62% of the work content saddling the targeted process (for example, sales invoice generation) is driven by the need to rework or research the defective or missing inputs passed on from the upstream operation. Accordingly, it would make sense to redirect the scope and strategy of the kaizen to encompass defect analysis, root cause identification, and verification to develop and implement the appropriate countermeasures.

If proper kaizen scope by analogy is like a laser beam, scope creep represents the floodlight . . .

high wattage, low focus, and ultimately little in the way of results. There are three basic drivers of scope creep:

1. the persistent introduction of side agendas by team members,
2. uncontained excitement over new data and insights obtained during the observation of reality, and
3. executive attention deficit disorder (good luck with that!).

The first two drivers can be addressed through the diligent re-grounding of the team by referencing the event area profile and the kaizen targets along with the use of a simple "parking lot." The parking lot (see Figure 6-20), typically a flip chart, is a tool that enables team members to record issues, problems, or opportunities that are either out of the event scope or well beyond the team's sphere of influence. By "parking" the item(s), the team acknowledges that the item(s) is outside of the scope, but because it is recorded, it will not be lost. With this comfort, team members can then move forward and remain true to the scope. The parking lot items should be briefly reviewed at the report-out and then with the appropriate level of management.

If the parking lot strategy fails to refocus a team member with a pet agenda, there should be a one-on-one meeting between the team leader and the team member to realign. Sometimes the problem is a lack of understanding on the part of the team member relative to the kaizen process (training issue?). Other times the issue may be more malignant and persistent, thereby requiring management intervention and possibly ultimate removal of that person from the kaizen.

Plan/Prioritize

Plan/prioritize activities surround the adherence and management of the general daily plan and the kaizen team's specific plan. Both plans are critical. While the general daily plan is dictated by the kaizen schedule-at-a-glance (see Table 5-2), the team-specific plan is really the time-phased steps. The steps are reflective of the team's strategy and the big picture view of the detailed prioritized activities and countermeasures.

> **Plus Delta**
>
> Every day, with the exception of the first, we kick off the kaizen day with a "plus delta" activity. During the activity, team members individually reflect for several minutes and then record on sticky notes single, brief thoughts relative to what was positive about the process (plus) during the last 24 hours and what should change or be improved upon (delta). Team members then attach the plus notes on the left side of a flip chart and the delta notes on the right side. There are several objectives to this practice. The first is to try and catch problems while we can still do something to improve the event. But, perhaps more importantly, we are always talking about and coaching lean leadership behaviors. Having the coach (sensei) which, in this case, is the leader of the team, stand up and critically examine his own process is key to leading by example. We want our leaders to constantly seek opportunities to eliminate waste and add more value for the customer. As coaches we can't just talk about what others should do. We demonstrate it in our own leadership behaviors as a learning tool for others.
>
> What is amazing to me is the range of input that you get from using this tool even after having run similar events hundreds of times. I have seen some groups simply suggest that the food is not good and the room is too cold. Then I've seen other groups get in-depth about value stream mapping techniques and how they just didn't see the connection to their own world. Others have suggested that the training approach is not working in their case. These are all opportunities to train the team on how to focus on the process and do simple root cause analysis.
>
> The bottom line is that each time there is a sense of things continually getting better through process improvement . . . this, after all, is the essence of lean. It is not about perfection; it is about continual improvement!
>
> *Technique shared by Joe Murli*

Work hours are generally defined by the kaizen week schedule and should be closely followed, unless the team agrees that it should work additional hours or that it needs to flex the schedule (for example, to conduct needed observations on the third shift). Typically, such a schedule flex can be accommodated by one or two team members for a day or a portion of a day. This should never

> **Executive A.D.D.**
>
> Some executives view kaizen teams as a personal special weapons and tactics (SWAT) team that can be instantaneously redirected and refocused. This ineffective behavior is often caused by: 1) a lack of understanding of or disregard for the kaizen process, and 2) limited familiarity with the gemba. As a result, kaizen scopes can quickly expand and shift. This bastardizes the kaizen process, frustrates the team members, and forces them to go a mile wide and an inch deep. These same executives often then complain that the event was not effective. There are several potential countermeasures to address executive attention deficit disorder (A.D.D.): training, kaizen team participation, team leader meeting participation, coaching from a superior . . . and prayer.

preclude scheduled items such as the daily team leader meeting.

If workers are normally allowed and subject to standard break times and break durations, these should be respected during the kaizen by *all* team members, unless there is a team consensus to modify the policy during the kaizen. There should be no double standards. No one should compel others to work through breaks; this only breeds contempt and is counter-productive. Often, as the kaizen progresses, team members become more and more engaged and will flex their work time/breaks to accomplish the assigned tasks and achieve kaizen targets.

The team-specific plan should be articulated and recorded on a flip chart. It reflects at a summary level the big picture things that must be accomplished during the day, including the time of the day, for example: "1) plus/delta review (7:30 a.m.), 2) 'trystorm'/design/build new fixtures and point-of-use storage (by 10:30 a.m.), 3) try/develop new standard work sequence (by 2:00 p.m.), 4) begin to observe/document new standard work (by 6:00 p.m.)," etc. The plan must include assignments and status and, to that end, be reviewed four or more times per day, and supplemented by the more detailed to-do list or countermeasures.

The first day of the kaizen is usually structured (kick-off meeting, training, validation of initial strategy, documenting of reality, etc.). The event should then settle into a rhythm or cycle of daily

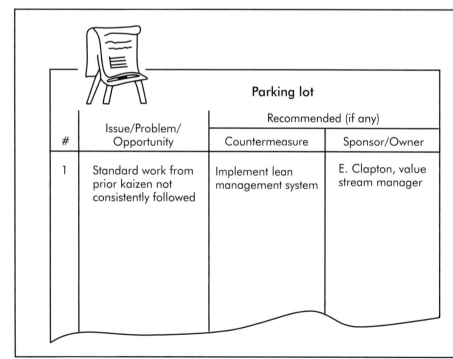

Figure 6-20. Parking lot.

Gemba Tales

Kaizen day-by-the-hour . . . for a deeper purpose. I cannot remember which of the Shingijutsu consultants hammered me, but the message stuck. He was running multiple events and required hour-by-hour, plan-to-actual results from each kaizen team. These were posted on a flip chart at the cell where the event was being held. Plans were made for half-day increments. Team leader requirements for the team leaders' meeting included presenting the plan-to-actual results for the past 24 hours and plans for the next half-day. If a plan was not made in the hour planned, the team leader had to show the countermeasure(s) taken to improve the performance and avoid a repeat failure without additional "leadership" intervention.

Feeling entitled to ask, "Why?" I did! Well . . . in the normal manner, I was first admonished and then "volunteered" to spend the next 3 days with the sensei. Every 2 hours we walked the gemba and stopped at each team's flip chart. At the right moment (after being humbled multiple times by the teams that did not make a plan and hearing their excuses of why, not what they were doing to make it better), the sensei finally told me "why."

The sensei explained, "It was so that I will know that when I return there will be no excuses as to why the cell is not performing to the new standard work at the standard work rate. I will know that you and your people can make a good plan and execute the plan. If an abnormal condition appears, it will be addressed quickly so that next hour they will make the plan!"

He was driving us to move away from, "Did we make the month?" to "Did we make the week, day, shift, or hour?"

Performance of plan-to-actual results on an hourly basis became one of the vital few metrics in the "pay for performance" compensation plan developed at Brooks Electronics (Wiremold). It was pretty successful. More importantly, the "hourly culture" really drove customer service, productivity, and quality within the cells.

Tale shared by Richard A. Jeffrey

---

planning. This cycle for team-level planning starts and resets at the conclusion of the day ("sunset meeting") with the "plan for tomorrow," before the team members go home. This timing affords the team leader to share any insights or redirection that may have been provided during the team leader meeting.

The "plan for tomorrow" is briefly reviewed again with the team first thing in the morning to refresh, refocus, and instill the appropriate sense of urgency. The specific plan is implicitly and explicitly comprised and directed by:

- alignment and performance versus the kaizen targets (reviewed with the team as part of the sunset meeting),
- current strategy,
- prioritized countermeasures and their completion status, and
- plus delta activity feedback.

At mid-morning and mid-afternoon, just prior to the team leader meeting, the team leader should review the status of the daily plan with the team, along with each of the countermeasures and direct/redirect as required (unaddressed issues and barriers will be communicated to leadership at the team leader meeting and hopefully resolved). At the mid-afternoon review, the team leader should also update the kaizen target sheet, reflecting the end-of-day performance versus the targets and formulate the plan for tomorrow. Both of these are agenda items for the imminent team leader meeting.

During the team leader meeting, the team leader will receive validation or redirection of the plan for the next day. Other relevant feedback will be provided. At the end of the day, after the team leader meeting, the team leader will conduct the sunset meeting, reviewing the status of the current day's plan, the countermeasures, and the plan for tomorrow.

Assign/Execute

The assign/execute activities are an extension or continuation of plan/prioritize. It is within this step that the team leader must gage the work content and degree of difficulty as well as any assignment sequencing issues (for example, team manager standard work cannot be finalized until the visual controls are substantially implemented). Further, the team leader must match assignments to the capabilities of the assignee(s)

>
> **Gemba Tales**
>
> **Never look a quick 7% productivity improvement opportunity in the mouth.** Our kaizen team was focused on developing standard work to improve productivity in two different work centers. One team member, "Greg," a young engineer, was resisting some of the proposed countermeasures. The team leader requested that Greg share his concerns with the team. The engineer thought the team was moving prematurely. He believed that they should first quantify everything and then rigorously validate, through process modeling, that the improvement ideas would indeed be successful. Notwithstanding his desire for such precision, Greg added, "Chasing after 10–11 seconds of improvement is a waste of time anyway." The team leader responded by demonstrating that 10–11 seconds of improvement could be realized with a quick low-cost/no-cost option. Further, he explained that this apparently inconsequential savings translated into a 7% productivity improvement and the capture of $500k in new sales that were previously inaccessible due to capacity constraints. Two morals of the story: 1) seconds count and 2) get done what you can get done *now*!
>
> *Tale shared by Edward P. Beran*

while considering the urgency and strategic importance to the success of the kaizen. And, as if those considerations are not enough, the team leader must anticipate and address "social loafing" situations in which team members may be less than inclined to work as diligently as they should.

Depending upon the dynamics, the team leader may assign two or more people to a single countermeasure. The "divide and conquer" approach, in which the team members separate and attack different assigned countermeasures, is one of the most effective strategies for making high-impact improvements quickly. Another important consideration is to ensure that the team members do not get bogged down by over-engineering elegant improvements. Quick and dirty is better than slow and fancy. Implicit in the "quick and dirty" approach is the opportunity to quickly engage in a "trystorm," learn and repeat cycle, ultimately bringing forth a more effective solution to the problem at hand.

When making assignments, the team leader should be mindful that it is a good thing to provide leadership participants (company or union) with an opportunity to "roll up their sleeves." Team leaders should have no hesitation in making people in leadership positions perform physical work. As part of a change management strategy, it is important for others, whether on the kaizen team or not, to see their leaders fully engaged and contributing sweat equity.

Track/Status

Track/status of activities continues the natural progression from the plan/prioritize and assign/execute activities. The subject matters of tracking and status evaluation are the assigned countermeasures, daily plans, and performance versus kaizen targets. It is all about focus, accountability, and instilling a proper sense of urgency so that plans and assignments are executed and, most importantly, kaizen targets achieved. Implicit in this is the need to easily flag abnormal conditions in as close to a real-time manner as possible. Abnormal conditions represent daily plan items and/or countermeasures that are anticipated to be accomplished later than the assigned date (and hour) or that are already past due, or the existence of real or anticipated gaps in the performance versus kaizen targets. Examples of these situations include:

- There is evidence that one element of the daily plan (for example, completion of time observations by noon) will not occur.
- The countermeasure for building a prototype fixture will be at least 24 hours past due.
- The kaizen targets call for, among other things, the implementation of at least one safety improvement per day, and it is day 3 and the team has implemented only one safety improvement thus far.

Consistent with lean, the desire is to have the kaizen team environment as self-explaining, team managed, and self-correcting as possible. The explicit nature of planning, prioritizing, assigning, tracking, and "statusing," along with the frequent review and visual nature, should make it evident to all team members (and any casual observer, for that matter) what needs to be accomplished, by when, by whom, and whether the item is on track. As such, team members can respond to abnormal conditions by refocusing, employing a different approach (for example, a simpler, more producible prototype fixture), or seeking the assistance of the team leader, teammates, sensei, and possibly, by means of the team leader meeting, management.

The most effective tracking and "statusing" tools are simple and visual. The "status pie" in Figure 6-21 satisfies both attributes. In addition to being used on countermeasure sheets, it can also be applied to daily plan sheets. The pie, which is sectioned into quarters, easily displays assignment status. Pie pieces should be colored only by the leader or sensei as each item progresses and/or as part of the frequent daily team update meetings. The restriction on pie "filling" is because assignees often are too generous in determining

> ### Cross-talk
> Often during multi-team kaizen events, one or more teams end up working on scopes that have points of intersection. For example, one team may be working on rationalizing a whole slew of productivity reports into one actionable and simple report, while another may be working on a simple, real-time visual control for the purpose of tracking file investigation strategy and status. In such a situation, there is an opportunity for interdependence/synergy. Without communication and coordination, there is a risk of misalignment and muda. Accordingly, each team should maintain a "cross-talk" flip chart where potential synergies or issues are recorded. The respective team leaders can visit each other several times a day, review the flip charts together, and make sure their teams are in lock-step.

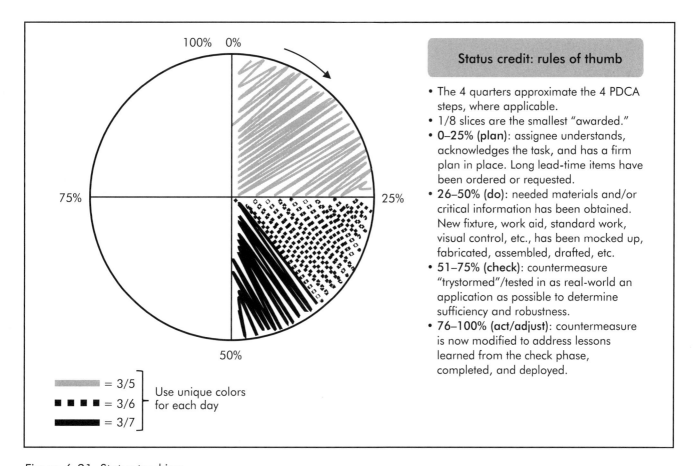

Figure 6-21. Status tracking.

the percentage of completion and because autonomous filling often means that the team leader will be unaware of the changing status of tasks and thus may make erroneous strategic or tactical assumptions. Pie status should be recorded in colored markers, with a different color used each day, and reflected in a key so that daily progress can be easily discerned.

Another simple and visual tool for tracking the status of activities is the kaizen target sheet, as shown in Figure 6-22. As discussed earlier, this sheet must be updated on a daily basis with end-of-day status reflected in the appropriate daily column. It is a critical work strategy tool and provides critical input for the team leader meeting. In fact, if the team leaders were limited to only one piece of paper, it would be the kaizen target sheet.

Document/Visually Manage

The document/visually manage activities are foundational to the other work strategy activities. They are wholly consistent with the following core elements of lean:

- 5S and visual controls,
- flow,
- pull, and
- standard work.

From a work strategy/administrative process perspective, a kaizen event that does not satisfy these elements is not going to achieve its full potential. Just as in any operation, the process should "speak to us," meaning that visually, anyone can quickly tell what the process is, whether or not it is working or in control, and its status.

An all-important first step for any event and in any team break-out room is a modicum of 5S. It does not take long for a break-out room to get out of control with supplies, papers, food and food containers, etc. While there probably will not be any scrubbing (the third "S"), the following S's do apply:

- *Sort*—get rid of excess materials, supplies, information, etc., on the floor, table, or walls.
- *Straighten*—there is a place for everything and everything is put in its place. This is where an organized, visually controlled team supply kit is critical.
- *Standardize*—a diligent team leader, co-team leader, or designee can keep the team mindful and disciplined enough to maintain the first two S's for the duration of the event.

Good kaizen documentation and visual management uses "storyboarding" as a tool. This facilitates visual controls and flow, and provides valuable insight into whether the team is following standard work and whether its activities are "pulled" by the data. If the team's break-out room has sufficient wall space, the team leader should storyboard the storyline on the wall. If wall space is minimal, then use the windows. If that is not sufficient, often the use of a large piece of corrugated cardboard can serve as a temporary wall(s). It does not have to be beautiful, just effective.

Storyboarding requires the taping of forms and flip chart pages on the wall in a progression that reflects the sequence of the scientific kaizen process. For example, the first item (whether moving clockwise or counterclockwise, it does not really matter) should be the profile form, followed by the (daily) updated kaizen target form, followed by the detailed takt time calculation, preferably on a flip chart, followed by a flip chart reflecting the initial team strategy, etc. Such a display, which can be supplemented by a flip chart of the storyline element sequence, quickly communicates to participants and observers the flow and progression of the kaizen event. It facilitates a common team understanding of the "big picture" and provides insight into the team's logic. Storyboarding also highlights whether the forms are being used properly, reveals whether the team is following standard work, and provides quick insight into the status of countermeasures. Further, it readies the team leader for the daily team leader meetings and positions the team to make a rather effortless preparation for the report-out.

The storyboard will also make it evident as to whether or not the team members let the data, reflective of the pre-kaizen situation, lead them in terms of gathering additional information, graphically portraying and analyzing it, forming

Kaizen Event Target Sheet

Location: Ellington
Event: Whirlpool Assembly
Team lead: John Quincy

Team name/#: EL-#04-5/YY
Date: May 14–May 18

#	Key Measurement	Start	Target	Day 1	Day 2	Day 3	Day 4	End	IMPACT	
									Difference	% Improvement
1	5S score									
2	Productivity	8.3/shift	10	8.3	8.3	11.3				
3	Throughput	83	90	83	83	90.4				
4	Crew size	10	8	10	10	8				
5	Lead time	1 Day	Same day	1	1	Same day				
6	Cycle time (seconds)	n/a	1,710	1,928	Test	1,410				
7	Setup/changeover time									
8	Defects									
9	Scrap									
10	Space (square feet, square meters)	1,875	900	1,875	625	625				
11	Inventory									
12	Work in process									
13	Parts' travel distance (feet, meters)	2,752	50	TBD	TBD	35				
14	Walking distance (feet, meters)	2,008	100	TBD	TBD	66				
15	Other:									

Figure 6-22. Kaizen target sheet: daily status. (For a blank form, see Appendix A.)

countermeasures and strategy, etc. This is "pull" as opposed to a situation where a team "pushes" its preconceived "improvements," which are often inconsistent with the demands of current reality.

REPORT-OUT

The final presentation, referred to as the "report-out" or "out-brief," represents a 20-minute briefing that is sandwiched between other agenda items. The presentation is made by *all* members of the kaizen team to an "audience" for the purpose of communicating the what, why, who, when, where, and possibly, most importantly, the how of the kaizen event. It is a time for the team to share and be recognized for its accomplishments.

Essentially, the report-out walks the audience through the kaizen storyline, reflecting the sequence of the scientific process inherent in kaizen—how the team used different tools to identify the waste, then acknowledge the waste and, ultimately, eliminate it. Quite often, the report-out process provides a *eureka* moment for the first-time kaizen participant, when he sees it all come together within the now completed storyline. Previously, it may have been camouflaged by the flurry of activity and high intensity of the kaizen event.

Kaizen is about meaningful improvements, not superficiality or apparent effectiveness. The time devoted to presentation preparation represents time not spent on real kaizen. So, while a crisp presentation is desirable, it should not be overdone.

Presuming that the team stayed "on path" throughout the kaizen, as guided by the standard work reflected in this book, keeping to the rigor of standard forms, kick-off and team leader meeting feedback and direction, and facilitator and sensei coaching, the presentation preparation time should be minimal. In fact, the full-team focused preparation time, including rehearsal, should not exceed 2 to 3 (team) hours.

Preparation should be directed by the team leader and centered principally around:

- team member assignment of the various presentation segments;
- final assembly of the kaizen storyline, which should have been storyboarded on the breakout room wall throughout the event;
- transferring small documents (for example, the kaizen target sheet, improvement idea forms, etc.) to overhead slides (acetates or "pasted" in presentation software) or, if they were not initially recorded on a flip chart, enlarging by copying them on flip chart paper;
- readying of "show and tell" items (for example, old and new fixtures), before-and-after photos or videos on a projector, etc.; and
- conducting a rehearsal or two.

See Figure 6-23 for guidance on presentation elements, sequence, and cycle time.

It is important to remember the purpose of the final presentation. It is *not* to torture the team members, many of whom are not experienced public speakers and dread such situations. The team leader should assign presentation segments accordingly, seeking to give team members portions that play to their strengths. For example, it is obviously best for a team member to present on what they worked with most intimately during the kaizen.

In situations where the team member is not fluent in the primary language of the audience, it may be best to assign a segment that is high in visual content—before-and-after photos, spaghetti charts, etc. Good visuals allow the opportunity for the presenter to do color commentary rather than "play by play." No matter the language capability of the presenter, verbosity is not a virtue. The

---

Measurement of Success

While the kaizen event area profile and the kaizen target sheet specifically reflect the targets and measures, the report-out should explicitly state how workers and management can easily measure and track success without having to periodically engage an analyst to apply "voodoo math" to figure it out. In other words, the key measures have to be few in number, simple, timely, and visual so everyone can tell if the post-kaizen situation is performing as advertised. Better yet, the worker-managed measurements should be incorporated within the lean management system.

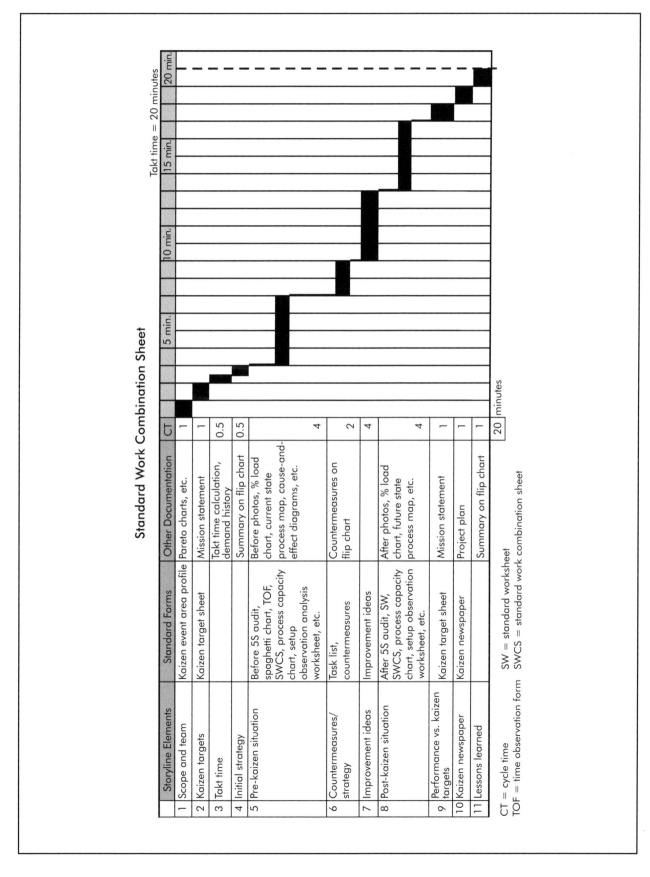

Figure 6-23. Team presentation: standard work combination sheet.

> **Gemba Tales**
>
> **Bluto?** By design, not much time is invested in preparing for the final report-out. However, presentation role clarity (what each person is presenting), review of key ground rules, and a quick rehearsal are valuable.
>
> During one report-out a team member who was part of the maintenance department was slated to quickly review the lessons learned. What came out was far different. He slipped into the role of "Bluto" from the movie *Animal House*. Specifically, the scene where Bluto, played by the late Jim Belushi, is trying to rally his fraternity brothers by recalling how the U.S. came back, even after the "Germans bombed Pearl Harbor." Yes, the team member was on a roll . . . a wholly misguided one in which he recapped the entire kaizen event, gave recognition to individuals, and shared personal reflections on the event. Entertaining, yes, it was; appropriate or value added, no.

location of the report-out to team members and the invitees, securing and preparing the presentation room, including flip charts, projectors, seating arrangements, etc. The sequence of the team presentations (if there are multiple teams) is determined by the kaizen promotion officer. At the report-out, he also acts as the emcee, welcoming the audience, sharing the agenda, reviewing and enforcing the related report-out guidelines, introducing the different teams, facilitating the brief question and answer sessions after each presentation, introducing the sensei and senior manager, and ensuring that the report-out is conducted in an orderly and on-time fashion. Figure 6-24 reflects the report-out agenda and guidelines which, when followed, will facilitate an effective report-out process.

The audience is typically minimally comprised of:

- the site leadership team,
- outside visitors,
- other kaizen teams who worked concurrently with the presenting team throughout the event period, and
- facilitators and sensei.

presenter should not feel compelled, for example, to review every lesson learned. Two or three will suffice. The rest the audience can read (another reason for big and simple charts and slides). One learned sensei often said, "Charts talk, people don't."

The kaizen promotion officer and designees are responsible for communicating the time and

If the kaizen team(s) is plant-related in scope and the plant is unionized, it is appropriate to invite top (local) union officials to the report-out.

Agenda

1. Welcome
2. Review agenda
3. Review guidelines
4. Team presentation(s)
5. Brief question and answer session
6. Consultant comments/critique
7. Senior manager closing remarks
8. Recognition and celebration
9. Gemba tour (if desired and if time)

Guidelines

- Each team presentation will be conducted within a 20-minute takt time and will stick to the script (no commentary).
- Questions from the audience are taken only after each team has completed its presentation.
- Applause is appreciated after each team report-out.
- Sensei comments are comments and not a dialogue or debate. Clarifying questions may be asked.

Figure 6-24. Report-out agenda and guidelines.

> **Report-out Hijackers**
>
> Report-outs should begin with a brief review of the agenda and report-out guidelines. Within the guidelines is the very clear direction to hold questions until the teams have completed their presentations. Despite that clarity, executives often cannot contain themselves and begin firing off questions. Possibly this display of incontinence is because they really need to know the answer immediately or they want to show everyone just how smart they are. If nothing else, this behavior demonstrates a lack of humility and respect for the kaizen team members.

Clearly there should be some level of orientation for the union officials, not dissimilar to what management is exposed to (classroom training, kaizen event participation, benchmark visits, etc.). This will facilitate understanding and support.

If it does not jeopardize satisfying customer requirements, does not induce a hardship on employees and the presentation venue can accommodate a larger audience, it is appropriate to also invite:

- the employees whose jobs have been/will be impacted by the improvements made during the kaizen (although if the kaizen is executed effectively, adequate communication and orientation should have already taken place along with the prove-out of the new standard work, which will be supplemented with post-kaizen support);
- key kaizen support personnel, such as maintenance, who were instrumental in making the products of team brainstorming and "trystorming" reality; and
- those who would benefit from seeing the technology, energy, and potential of kaizen from a training or change management perspective.

It is desirable after the report-out for the team members to conduct a tour of the gemba to physically show and demonstrate the improvements to others. Visual controls, including standard work, should be posted as appropriate. Even more desirable, assuming it is appropriate from an environmental health and safety perspective and that the presentation would be audible to the audience, is to have the entire report-out at the gemba. This forces the executives to go to the gemba and enables the team(s) to really show off.

RECOGNITION AND CELEBRATION

The recognition of team effort and accomplishments, when warranted, starts as early as the first day of the kaizen. Like many things, there are formal and informal opportunities. Effective lean leaders take advantage of both.

Each team leader meeting, occurring on all but the last day of the event, presents a formal venue for leaders to express their thanks and give praise. This can be expressed verbally and/or take the form of simple but universally appreciated applause from management after a team leader delivers a cogent debrief that reflects sound lean understanding, good strategy, tactics, action, and results.

The daily lunches, typically provided to each participant, are another opportunity to recognize the teams. However, no one should underestimate

> **Value Stream Analysis Report-outs**
>
> Typically, the following items are covered in a value stream analysis report-out:
> - scope and team (kaizen event area profile);
> - product family analysis matrix and related takt time calculation;
> - future state performance targets as dictated by strategic imperatives—strategy deployment, competitive market assessment, etc.;
> - current state value stream map and related themes/issues;
> - future state value stream map;
> - current versus future state comparison of summary statistics, including number of steps, lead time, process time, roll-throughput yield, etc.;
> - improvement ideas;
> - value stream improvement plan;
> - next kaizen (it is helpful to have a first draft of the kaizen event area profile already prepared); and
> - lessons learned.

the impromptu visit from the interested ("Tell me/show me what you are working on. . . ") and supportive ("Do you have everything you need to accomplish your objectives?") executive. Such a visit should catch the team members in the break-out room and on the gemba as they conduct the kaizen. The visit should impart the message that leadership:

- recognizes the importance of the kaizen objectives,
- understands and expects a certain level of effort and sacrifice from team members and for that is appreciative,
- cares about what the team is doing and learning, and
- will ensure that the team has sufficient and timely support to achieve its targets.

Certainly, the report-out presents a grand opportunity for recognition. This final-day activity represents a time for each team to share its accomplishments. At the conclusion of each team's presentation, there is a brief question and answer session, followed by applause. Once all teams have delivered their presentations and the sensei has provided comments, then it is customary for the senior leader(s) to make a brief statement. This statement, among other things, should recognize the accomplishments of each team. As part of this concluding element of the final presentation or immediately after, it is appropriate to celebrate with a little more vigor by recognizing team members individually and awarding them a small token of appreciation. If any of the kaizen teams have union members, it may be appropriate to have top local union officials participate in the celebration in a meaningful way.

The award can be as simple as a certificate of appreciation, signed by one or more senior executives, or a company shirt or hat printed/embroidered with a reference to the kaizen/location/date, or a kaizen pin, etc. (In the interest of frugality, many companies present a shirt to each first-time kaizen participant and then award team members something less costly after each subsequent event.) While this type of recognition activity is more private than public (at least until the team members start wearing their new shirts), it is proper and fitting to more publicly acknowledge the team's accomplishments.

Public recognition often comes in the form of articles in the company newsletter and/or posting the brief one- to two-page kaizen summaries (in an A3 report-type format) in a prominent location at the site where the kaizen was conducted. The articles and posted summaries should, at a minimum, detail the team (with photo), the target, scope, improvement ideas, and accomplishments (see Chapter 7 for further information).

Often companies allow the hourly participants to take off the rest of the last day of the kaizen. This is in recognition of the long hours that they invested during the event and their accomplishments. However, no one should be dismissed until the report-out, recognition session, and any critical wrap-up action items have been completed (for example, final postings of standard work, finalization of the kaizen newspaper, etc.).

Finally, one of the most simple, but powerful means to recognize the efforts of a kaizen participant is to mail, directly to their home, a note card from a respected senior executive with a brief, handwritten sentence or two of thanks. This transcends what often may be seen as corporate propaganda, and is something that an employee can share with his or her family.

SUMMARY

- Event execution is the third phase of kaizen event standard work.
- Seven steps comprise the kaizen event sequence.
    1. The kick-off meeting is a 30- to 50-minute team meeting to formally launch the event during which the team leaders review their event area profiles, target sheets, and share their initial strategy. Senior management and the sensei also make remarks.
    2. Pre-event training—this brief just-in-time training covers kaizen methodology and lean topics that are often tailored to the individual mission of each team.
    3. The progression of the kaizen storyline, during which each team follows a specific

11-step PDCA-based plotline or rigor, often supplemented by standard forms and templates:

    i. scope and team,
    ii. kaizen targets,
    iii. takt time (customer requirements),
    iv. initial strategy,
    v. pre-kaizen situation,
    vi. countermeasures and revised strategy,
    vii. improvement ideas,
    viii. post-kaizen situation,
    ix. performance vs. targets,
    x. kaizen newspaper, and
    xi. lessons learned.

4. The team leader meeting process during the event represents a daily checkpoint process in which team leaders review the progress, plans, issues, and barriers with leadership. Lean leaders encourage support, challenge, and teach within this forum.

5. The kaizen work strategy comprises the approach and techniques, such as visual storyboarding, plus/delta reviews and frequent countermeasure status checks, by which team leaders and facilitators can enhance a team's event effectiveness.

6. The report-out is a 20-minute final briefing of each team's event accomplishments and follows the kaizen storyline.

7. Recognition and celebration conveys formal acknowledgement of each team's effort and contribution to the organization.

REFERENCES

Csikszentmihalyi, Mihaly. 1990. *Flow: The Psychology of Optimal Experience*. New York: Harper & Row.

Encarta World English Dictionary. 2009. Microsoft Corporation. Accessed from Internet 5-13-09, http://encarta.msn.com/dictionary_/countermeasure.html.

Imai, Masaaki. 1999. *Gemba Kaizen*. San Francisco, CA: Berrett-Koehler Communications, pp. 58–59.

Merriam-Webster OnLine. 2009. Accessed 5/12/09, http://www.merriam-webster.com/dictionary/storyboard.

Stolovitch, Harold D. and Keeps, Erica J. 2006. *Telling Ain't Training*. Alexandria, VA: American Society for Training & Development, p. 47.

7
FOLLOW-THROUGH

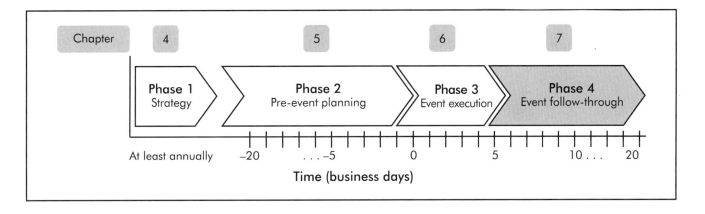

"He conquers who endures."
—Persius

PHASE 4: EVENT FOLLOW-THROUGH

In virtually every sport, part of the athlete's standard work includes follow-through. It is extremely critical whenever an athlete throws, kicks, shoots, hits, swings, bowls . . . you get the point. The athlete's follow-through facilitates repeatability, accuracy, and power. It also reduces strain (muri), thus enhancing longevity and sustainability. Kaizen follow-through is equally important for many of the same reasons, but it is often given short shrift.

At the conclusion of the final report-out, the kaizen team members often "feel" that they are done. The adrenaline of the report-out is followed by an afterglow with roughly equal portions of fatigue and relief. The team members have worked long hours during the event, probably pushing and accomplishing much more than they ever imagined. They have weathered the real or imagined stress of final presentation preparation and delivery—this milestone being the visible and formal climax of the event. Their perception is one of "mission accomplished" or, "The party is over, now who (other than me) is going to clean the dishes?" However, this is not the end of their focused improvement activities; it is just the conclusion of the event execution phase and the beginning of the event follow-through phase.

Indeed, without a deliberate and formal transition between the event execution and event follow-through phases, there is real risk of unsustained and unrealized improvements, and fuel for the lean nay-sayers. In other words, the team's effort during the kaizen, the possible disruption in operations, travel costs of the out-of-town participants and guests, purchased materials and supplies, overhead and overtime related to the support services, sensei fees, etc.,

Lean Leaders' Short "Must Do, Cannot Fail" List
- Demand and verify that event follow-through standard work is followed.
- Do not engage in serial or superficial kaizen.
- Personally audit (go to the gemba) the portions of the lean management system that employ the new leader standard work and supporting visual controls.
- Understand and immediately address the root cause(s) of backsliding.
- Whenever appropriate, require kaizen gains to be quantified relative to their financial impact.

will end up being an extravagant example of feel-good muda!

Post-event Follow-through Planning Session

Preparation for the post-event follow-up really starts, or at least is lined up, well before the final report-out. This timing is driven by pragmatism. Teams have a way of scattering almost immediately after the applause following the final presentation. Accordingly, post-kaizen event follow-through expectations must be established with the team leaders and team members from the very beginning. It is prudent to include a kaizen follow-through planning session as the last item within the kaizen event schedule at-a-glance (see Table 5-2).

The session, the agenda for which is reflected in Figure 7-1, should take no more than 20 to 30 minutes and serves as a primer for addressing the four basic follow-through categories or themes:

1. sustainability,
2. event management improvement,
3. communication, and
4. record retention.

These categories, their purposes, and elements are reflected in Figure 7-2. Consistent with lean, a checklist can help facilitate follow-through standard work and, ultimately, execution. The kaizen promotion officer should share the checklist with each team leader during the planning phase and revisit/re-emphasize it during the kaizen event so that they can anticipate and plan. See Figure 7-3 for an example checklist.

The kaizen promotion officer (usually without the benefit of the sensei who by now is probably en route to the airport) should conduct the post-event meeting immediately after the report-out and any recognition and celebration activities. The necessary participants include the team leaders and co-leaders from each team, the value stream manager, department manager(s) who have responsibility for the kaizen target areas and, as available, other members from the site leadership team. Senior management participation helps reinforce the importance of the post-event follow-through. It also facilitates the decision-making process, especially when there are decisions pending relative to things like resource assignment and countermeasure selection and de-selection.

Sustainability

Clearly, the most important of the four follow-through themes is sustainability. The purpose is to both verify and hold the gains that were (measurably) achieved during the kaizen event, or at least anticipated (because they were not wholly measurable due to the lack of cycles to observe, the extremely long lead times, etc.).

Serial and Superficial Kaizen

Some leaders see kaizen events as if they are just another scan on a frequent-shopper reward card. They fail to understand that with each kaizen comes the responsibility to: 1) see it through to the completion of all newspaper items, and 2) ensure that the gains are sustained by integrating them within a successful and robust lean management system. A post-report-out litmus test occurs after each kaizen event. Associates will constantly judge the commitment and stamina of leaders to see if they "walk the walk." This is a test of leadership's credibility and competency, without which employees will neither risk change nor feel compelled to work to the new standard work. Even worse, employees will see the kaizen event as a charade.

Agenda	Actions/Considerations
1. Checklist overview	Conduct a quick walk-through of the post-event follow-up checklist to advise meeting participants of content, responsibilities, and expectations.
2. Finalize newspapers	• Each team leader presents a kaizen newspaper. • Meeting participants probe to ensure the kaizen newspaper is accurate, complete, timing is proper, etc.
3. Collect event evaluations	• Gather kaizen event evaluations. • If there is time, the kaizen promotion officer scans the evaluations for notable comments and trends, and shares them with meeting participants.
4. Conduct quick post-mortem on event	Drawing upon direct experience from this kaizen, evaluation feedback, lessons learned, and sensei comments, identify items to "start," "stop," and "continue" doing.
5. Formulate kaizen event management improvement countermeasures	Based upon above, identify countermeasures (what, who, by when).
6. Collect hard/electronic copies of key documents	Key documents include the kaizen event summary report, report-out package, and finalized newspaper.

Notes:
- Meeting is run by the kaizen promotion officer immediately after report-out in a break-out room
- 20- to 30-minute duration
- Participants include: team leaders, co-leaders, value stream manager/process manager, department managers for target areas, site leadership team (as available)
- If the participants have the requisite authority, this is a good forum to review the parking lot items and make final dispositions

Figure 7-1. Post-event meeting agenda.

Sustainability is largely driven by three major elements:

1. kaizen newspaper execution,
2. leader standard work, and
3. the post-kaizen audit.

The "kaizen newspaper" here serves as a euphemism for post-kaizen action items, whether limited to the items captured in the kaizen newspaper form itself or, as often happens, extending to more necessarily sophisticated transition planning. Leader standard work, one of the components of a lean management system (see Chapter 3), facilitates the day-to-day verification that the new system is operating as designed. This is primarily a means for determining process adherence or compliance. In contrast, the post-kaizen audit verifies performance as well as process compliance and newspaper execution.

Kaizen Newspaper

The reasonable assumption should be that the kaizen newspaper has been diligently developed prior to the report-out (see Chapter 6). That said, often the kaizen newspaper is thrown together in the heat of battle just before the report-out. Therefore, review of the kaizen newspaper in the

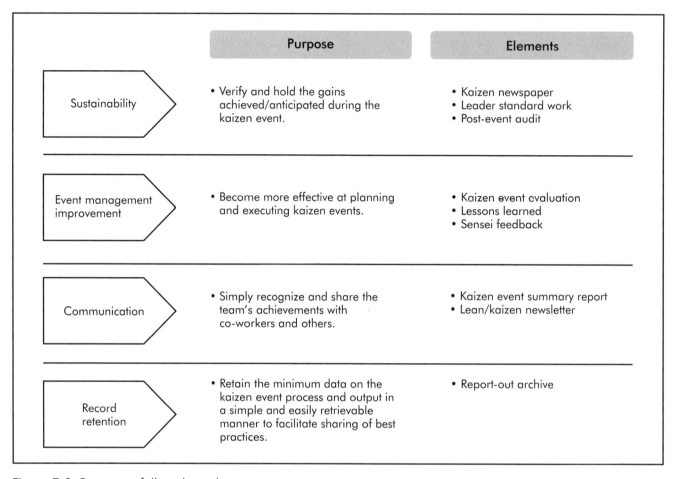

Figure 7-2. Post-event follow-through.

 Flow Kaizen Follow-through

The primary output of every value stream mapping event is the value stream improvement plan (VSIP). The VSIP represents a detailed, time-phased roadmap for transforming the particular product or service family's value stream into the leaner version reflected within the future state map. Accordingly, flow kaizen follow-through is largely about the formal and frequent (one to two times per month) review of the VSIP. This program review-type rigor holds people accountable and drives execution. As might be expected, during the ensuing months of VSIP implementation, the value stream leaders will gain further insight into the value stream, its opportunities, and challenges. Therefore, it is often appropriate to adjust or refresh the future state map and VSIP.

post-event planning session should be very deliberate. Due to the fact that an effective kaizen will likely generate only a handful of kaizen newspaper items, the review should not be onerous. The post-event planning session seeks to ensure:

- newspaper completeness and precision,
- pragmatism, and
- clear expectations relative to posting, review, and execution.

There is a risk that the newspaper may be incomplete (for example, no one is designated as the person responsible or there is no due date) or it may reflect vague countermeasures thereby leaving confusion as to what really needs to happen over the next few weeks. The most simple test is whether the newspaper addresses the "what,"

Chapter 7: Follow-through | Kaizen Event Fieldbook

Post-event Follow-through Checklist

Team: _____ Report-out date: _____ Process owner: _____ Kaizen promotion officer: _____
Team leader: _____

	Post-event Categories		Required Activities	By Whom	Final Day of Kaizen	Post-kaizen Period Week 1, Day: 1	2	3	4	5	Week 2	Week 3	Week 4	30-Day Audit
Sustainability	1. Kaizen newspaper	1.1	Finalize newspaper.	Team										
		1.2	Post newspaper.	TL										
		1.3	Status newspaper items.	TL, VSM, KPO										
	2. Leader standard work (LSW)	2.1	Complete final LSW draft and validate linkages to visual controls.	Team, KPO										
		2.2	Validate and approve LSW, allocate ownership to appropriate leaders, and integrate into existing LSW.	VSM, KPO, TL, supervisor										
		2.3	Deploy LSW ("go live").	VSM, supervisor										
		2.4	Audit LSW. Identify/deploy necessary countermeasures.	VSM, KPO										
	3. Audit	3.1	Determine current performance vs. reported kaizen achievements (including financial impact). Reconcile differences if any, assign countermeasures as required.	VSM, KPO, TL										
		3.2	Conduct review of checklist status (all of above items), issues, etc. Identify/deploy necessary countermeasures.	SLT, VSM, KPO, TL										
Event management improvement	4. Kaizen event evaluation	4.1	Complete and submit evaluations.	Team members										
		4.2	Review and compile submitted evaluations (also reference lessons learned). Conduct follow-up discussions with submitters if clarification or additional information is required.	KPO										
		4.3	Reference formal and informal sensei feedback.	KPO										
		4.4	Complete event post-mortem (start, stop, continue doing).	KPO										
		4.5	Identify necessary countermeasures. Review with appropriate people and assign.	KPO, others as required						When appropriate	When appropriate		
Communication	5. Kaizen summary	5.1	Complete kaizen event summary report.	TL										
		5.2	Post/distribute summary report.	KPO										
	6. Newsletter	6.1	Incorporate kaizen activity and results within regular lean/kaizen newsletter.	KPO						When appropriate	When appropriate		
Record retention	7. Archiving	7.1	Submit summary report-out package to KPO.*	TL										
		7.2	Quantify financial impact of improvements. Submit to KPO.	Finance and accounting support										
		7.3	Catologue and electronically file report-out package and financial impact estimate.	KPO										

KPO = kaizen promotion officer, SLT = site leadership team, TL = team leader, VSM = value stream manager, LSW = leader standard work
* Summary report-out package is to include at least the following: kaizen event area profile, kaizen target sheet (with results), kaizen newspaper, and kaizen lessons learned.

Figure 7-3. Post-event follow-through checklist. (For a blank form, see Appendix A.)

"who," and "when." If these criteria are missing, then there is little hope for execution.

While completeness and precision are important, the plan embodied within the newspaper or the project plan must be achievable. In other words, the assignees for the various tasks must have the capacity and the capability to get the task done by the due date. This considers not only the available time and skill sets of the responsible persons, but also their authority (if they need to direct others or purchase certain items) and the lead-time performance of key internal and external suppliers.

Once the newspaper is deemed to be accurate and complete, the attention must turn to execution. It is critical that the team leader understands that she is ultimately accountable for driving completion of the items. The kaizen newspaper must be posted in a central and visible area in the target location to facilitate frequent review and the flagging of any abnormal situations (past due countermeasures). If there is a lean management system in place, the tier I or II visual board, at which daily meetings are conducted, is an excellent standard location for newspaper posting. A good tiered meeting agenda should call for a weekly status review of kaizen newspaper items.

The newspaper should be updated on at least a weekly basis. Only the team leader or kaizen promotion officer has the authority to maintain the "% complete" field as progress is made and to record the "date complete." This restriction of the pen is necessary to ensure that any progress reflected on the newspaper is valid. Countermeasure assignees are sometimes gratuitous when assessing the level of their progress. When leaders regularly review the newspaper, they need to be confident that it is consistent with reality.

If the kaizen newspaper is comprised of a project plan for a high-intensity roll-out or transition, then it probably makes sense to conduct weekly or bi-weekly project reviews. In this instance, the project manager should conduct the review with key stakeholders and managers for the purpose of proactively managing the project and any related schedule, cost, technical, and/or change management risks.

Leader Standard Work

Kaizen events, by definition, improve the way things are done. Improvements require change, and thus within the scientific context of lean, new "systems" are developed and deployed. These systems are comprised of and reflected in things like standard work, new work area design, and visual controls.

The inherent steps of "trystorming" and validation during the kaizen event do not necessarily guarantee that the new or enhanced system will operate flawlessly going forward. Clearly, there will be challenges, some behavioral—workers may not consistently work in accordance with the new standard work, and some technical—fixtures, software, etc., may not work as intended. Whether anticipated or not, these abnormal challenges will manifest themselves over time.

Abnormal conditions require timely countermeasures, without which the expected kaizen gains will be unrealized and unsustainable, and new improvements elusive. Sustainability of kaizen gains is one of the most common issues

---

Gemba Tales

Preserved in plastic? As he should have, a vice president of operations postponed the forthcoming kaizen event because his folks had made scant progress on the kaizen newspaper items from the previous event. While this "punishment" was appropriate, the situation should not have festered as long as it had (in excess of 4 weeks). In fact, management, in some sort of bizarre lean archaeological preservation technique, had taken advantage of a recent water leak. They made use of clear plastic sheets to cover the prior month's kaizen storyboard and unfinished newspaper to protect them from water damage. Of course, this made the newspaper, among other things, inaccessible. No problem, they were ready to hang the next kaizen event work product over the old! An effective lean leader requires a weekly review of open kaizen newspaper items, preferably as part of a lean management system tiered meeting. This rigor drives accountability and action. Missed items beg immediate explanation and a plan for recovery.

that undermine the lean transformation journey. This is where the rigor of standardize-do-check-act (SDCA) is so critical.

Leader standard work (LSW) is an integral component of the lean management system (LMS)—an extremely powerful SDCA tool. LSW helps ensure that existing standard work and the related systems, whether newly installed or long adopted, are operating as intended. See Figure 7-4 for an example. LSW specifies aspects of daily management for people in leadership roles, including:

- *Where*—LSW indicates the location where a leader must physically or virtually be to observe and verify a prescribed condition (in situations where there are widely dispersed physical locations this may include intranet review of "electronic" visuals).
- *What*—LSW defines the normal desired condition as evidenced by simple "drive-by" visual controls. By defining normal, it is much easier to determine the abnormal and then implement appropriate and timely countermeasures.
- *When*—LSW prescribes the frequency at which the audits should be performed.

With each new kaizen, the LSW should be expanded to incorporate the new "where," "what," and "when" for the different leadership levels in the organization. (Unfortunately, for most companies this is new territory. They do not have a LMS to expand. In this situation, the best thing to do is start one, often through LMS training and an initial LMS kaizen.) Assuming there is a lean manufacturing system foundation, the typical thinking is to expand it later, sometime after the kaizen. This is a serious error. The obvious risk is that "later" never comes or does not come for a long time. Accordingly, the LSW must be drafted and the linkages to the visual controls validated before the kaizen is completed.

During the kaizen event, the team should identify the critical elements of the new "system" that they have implemented. This may include, for example, the implementation of new standard work for a specific process, the installation of a new, first-in-first-out (FIFO) lane with maximum levels for documents waiting for processing, etc. The team should ensure that these elements, exploded in critical detail and supported by appropriate visual controls, are captured within expanded leader standard work. Consistent with kaizen, the next generation LSW should be validated by at least several walk-throughs by team members and hopefully leaders of the target process. The team should not worry about how the LSW will cascade up (or down) to the various levels of leadership; the important thing is getting the where, what, and when down. Leadership can figure out the "who"; whose lean standard work specifies these audits?

Within 48 business hours after the kaizen, the team leader or kaizen promotion officer should review the updated leader standard work with the value stream manager and, if possible, all the other leaders whose standard work has been impacted. This is done to facilitate their understanding and gain approval (none of this should be a surprise as most, if not all, should have attended the various team leader meetings and the final report-out). Immediately subsequent to this, the new/newly updated leader standard work will be enacted.

As part of any comprehensive lean management system, the completed leader standard

Gemba Tales

Leader standard work only works if you use it. The sensei had heard secondhand and then observed firsthand that sustainability was not the strong suit of a particular plant. While facilitating a kaizen, he quickly drafted several lines of leader standard work to show the leadership team what leader standard work looked like and how it could be deployed immediately. The plant manager and staff swore "thus and so" that they would do it. The sensei's skepticism was validated a month later, when the prior kaizen gains had slipped and the plant manager reluctantly admitted that leader standard work had not been implemented. Several months later, while in the midst of a lean management system kaizen, the plant manager experienced a "eureka" moment. In all seriousness, he asked, "Why wasn't this done sooner?!"

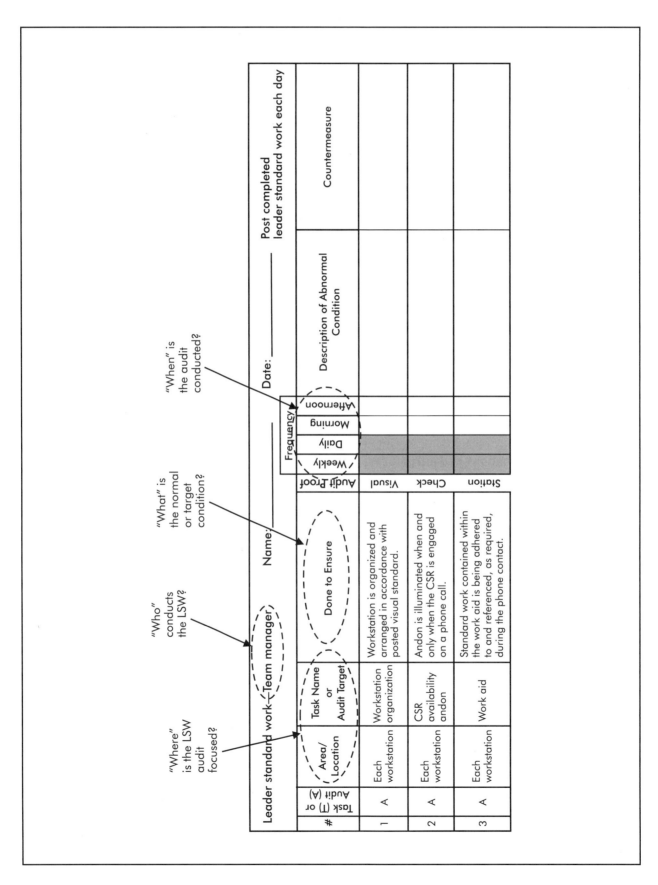

Figure 7-4. Leader standard work.

work should be submitted each day for review by the next leadership tier. Abnormal conditions will be the feedstock for daily tiered meeting discussions and countermeasure assignment. The kaizen team leader and kaizen promotion officer can easily conduct their own daily audit of their newly implemented "systems" by using the relevant portions of the leader standard work.

Post-event Audit

The post-event audit represents a comprehensive sustainability assessment and, barring serious issues, the formal and full transference of ownership to area management. Taking place on the event's one-month anniversary, this review is essentially comprised of the last required activities within the post-event follow-through checklist. An example agenda for the audit meeting is reflected in Figure 7-5.

The kaizen promotion officer, team leader, co-leader (if on site), kaizen event sponsor, value stream manager, department manager of the target area and, if appropriate, representatives from human resources and finance, participate in this review. The team leader and the department manager, with assistance from the kaizen promotion officer, prepare for, call, and lead the meeting. The kaizen promotion officer serves as the facilitator of the meeting, which is 30 minutes in duration.

The audit activities, with the exception of the financial review (which should be conducted in a private conference room), are conducted at the gemba. A quick review of the checklist status will "re-ground" everyone. This review is then followed by an audit of the:

1. kaizen newspaper status;
2. the completed and submitted leader standard work forms for the prior period, with special attention paid to recorded abnormal conditions and required countermeasures;
3. current performance levels as compared to the end-of-kaizen actual results reflected on the kaizen target sheet (for example, productivity, work in process [WIP] levels, etc.); and
4. calculated financial impact.

The specificity of the review and its gemba venue will quickly pierce any false pretenses relative to sustainability. In other words, if there

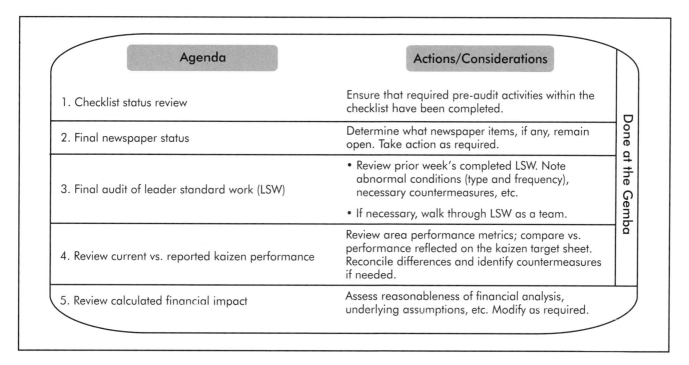

Figure 7-5. Post-event audit agenda.

are any significant problems, warning signs will be identified within the first few minutes of the audit. Abnormal conditions are addressed as required. Appropriate countermeasures may include reconstitution of the kaizen team, the formation of a new kaizen team to address upstream root causes, or a private coaching session between the senior leader and the responsible manager in which he or she is "re-educated" on the basics of lean leadership.

While the first four items of the audit agenda are straightforward, the last (financial impact) can be more challenging. Lean, in the end, is done for the benefit of the organization's stakeholders. At the risk of sounding mercenary, it is largely about financial performance. Granted this is the result of excellent customer service, least waste processes, engaged employees, etc., but there is no prosperity and certainly no employment security without corporate financial health. Many have found measuring the financial impact of lean, specifically kaizen events, elusive and, for the finance and accounting challenged, intimidating. The scope of this book does not extend to the vagaries of standard cost and absorption accounting, and activity based costing, and how they can drive the wrong behaviors. Nor does it encompass the pragmatism of value stream based profit-and-loss statements. Thus readers may want to explore the writings of Orest J. Fiume, Jean E. Cunningham, and Brian H. Maskell for further information. There are, however, two basic points to keep in mind:

1. Financial performance must be measured. It would be antithetical to lean at its most basic level (measure-improve-measure) to not measure financial impact. Further, while zealots understand that good process equals good outputs and thus well-executed lean will drive the numbers, demonstrating financial linkage is critical to any successful change management effort. Leaders should be able to quickly share an approximate lean return on investment (ROI) when asked.
2. The task does not need to be hard. Contrary to what many accounting folks espouse, precision is not required. Rounding to thousands of dollars is encouraged as it reflects the appropriate granularity of the calculations, while underscoring that the benefits are typically substantial enough not to get lost in the rounding. That said, it is important to engage a pragmatic person from finance and accounting (possibly get them to participate on the kaizen team itself) to help build simple models to calculate the benefits. This ensures consistency and credibility. Care must be taken to articulate key assumptions (show the math). Conservatism should be the rule, especially when it comes to "recognizing" the benefit.

The kaizen event benefits should mirror the performance versus the kaizen target, except they are obviously expressed in dollars. For example, the implementation of a raw material supermarket may effect a significant reduction in raw material inventory. A good supporting kanban-sizing calculation usually incorporates a comparison of the pre-kaizen inventory levels with new inventory targets (which, on average, may approximate half of the cycle stock, plus the buffer and safety stock). It is advisable to look at prior month inventory levels when determining the pre-kaizen inventory baseline. Often a multi-month average should be used. By comparing inventory level changes, part number by part number, and extending them versus the standard (or, better yet, actual) cost, it is relatively simple to determine the anticipated change in inventory investment levels.

Calculated kaizen benefits fall into a few basic measurable categories:

- margin enhancement,
- cost savings, and
- working capital improvements.

Of course, to understand the ROI, the expenses associated with the kaizen should be captured. These expenses include things like professional fees, kaizen promotion office costs, and travel expenses. See Figure 7-6 for further insight into calculating a kaizen event's financial impact.

EVENT MANAGEMENT IMPROVEMENT

An enterprise has to become extremely proficient at kaizen to have a chance at becoming

	Example kaizen drivers	Considerations
Margin enhancement	• Increased process capacity: setup reduction, total productive maintenance applications, etc. • Work content reduction: standard work, continuous flow, etc. • Yield/scrap improvement: mistake-proofing, standard work, etc. • Product simplification: production preparation process, value engineering, etc. • Space reduction: continuous flow, pull systems, etc.	• Incremental benefits of capacity improvements should only be recognized when they are known to be "sellable" or if they can result in the closure of an operation (cost savings). • Work content reductions/productivity should only be recognized when costs are truly eliminated (overtime reduction, labor reduced through attrition, etc.) • Space reduction benefits recognized if/when absorbed by another paying customer (department, lessee, etc.) or sold.
Cost savings	Work content reduction: report rationalization, standard work, value stream based organizational structure, continuous flow, etc.	Cost savings represent sustainable reduction in overhead or support function costs. (See above for further insight.)
Invested capital	• Inventory reduction: continuous flow, pull systems, etc. • Receivables reduction: standard work, visual controls, etc.	The application of material replenishment systems and lead time compression will drive down invested capital. Recognize benefits only when they have been demonstrated.
Kaizen cost	• Kaizen promotion officer salary and overhead • Travel costs • Professional fees	Capturing related costs will help gain insight into the return on a particular kaizen. However, return on investment is more meaningful when looking at the broader scope of multiple kaizen events over a period of time.

Figure 7-6. Kaizen event financial impact.

> **Gemba Tales**
>
> **Professional skepticism.** As part of the financial impact assessment of a recent and significant changeover reduction kaizen event, it was decided that the metrics should be included in the existing monthly reporting process. To facilitate this monthly task, pertinent operational data was tracked by the plant's financial lead person who then modified the monthly manufacturing report accordingly. This exercise allowed the monthly benefits to be routinely calculated in less than 15 minutes. A subsequent audit was conducted by the manager of manufacturing finance, who also just happened to be the supervisor of the plant's financial lead. The manager's knee-jerk reaction was predictable—heavily laced with a skepticism developed by witnessing years of "business folks" reporting outrageously exaggerated savings from their pet projects. He immediately insinuated that the kaizen benefit numbers were inflated. Of course, his perspective changed once the data capture and reporting processes were explained. The manager quickly realized that the numbers used were the same as those in existing monthly reports. Afterward, he commented that the benefits calculated "might even be conservative!"
>
> *Tale shared by Carl M. Cicerrella*

lean. Accordingly, to continuously improve, it is imperative that a company apply kaizen principles to the way it conducts events. This is a classic case of "physician, heal thyself," without which all credibility is lost.

In addition to the observational power of the kaizen promotion officer and other members of leadership, there are three sources of kaizen event management improvement opportunities:

1. kaizen event evaluations,
2. lessons learned, and
3. sensei feedback.

The event evaluation is a simple one-page questionnaire that each kaizen participant is asked to complete at the end of the event. The lessons learned are staples of the storyline. Sensei feedback is typically provided whether or not there is a willing audience and can be formal or informal, but always relatively blunt.

Kaizen Event Evaluation

The kaizen event evaluation form (see Figure 7-7) is given to each participant at the end of the event by the kaizen promotion officer. It is comprised of 10 straightforward questions. Question one seeks insight into the level of kaizen experience, two through five probe the effectiveness of pre-planning and kick-off, six and seven probe the sufficiency of event support, and eight through ten get at event success, desire to participate in future events, and opportunities to make the event better, respectively.

As with any evaluation or survey, the kaizen promotion officer looks for trends among the results, by event, by team, by experience, etc., as well as notes any significant outliers within the different populations. It may be beneficial to contact the participants to clarify or expound on their answers, although it may be prudent to provide the evaluator the option to remain anonymous. This may increase the probability of obtaining honest feedback, especially if there are executives involved in the kaizen. In no way should an individual's feedback be integrated into his or her personal performance evaluation. The form and the related process are wholly for the improvement of the company's kaizen event management.

Lessons Learned

Post-event reflections of the kaizen experience at both an individual and team level comprise the lessons learned. While the list of "lessons" is typically positive in nature, relating discoveries during the kaizen process, effectiveness of the team, and such, occasionally it contains constructive criticism. Often this highlights opportunities to improve future kaizen event management.

Example lessons learned topics include:

- Team composition—the team would have been more effective if the target process had more representation on the team.
- Pre-work—the initial strategy was understood before the kaizen. It may have been more efficient if, for example, in preparation for a kaizen within the insurance industry, the claim files that matched the target characteristics were identified prior to the event.

Kaizen Event Evaluation Form

Participant name (optional): _____ Team: _____ Event date(s): _____

1. Including this event, I have participated in:
   - ○ 1 Event
   - ○ 2 Events
   - ○ 3 Events
   - ○ >3 Events

2. I participated in this event because I:
   - ○ Volunteered
   - ○ Was assigned

3. I was informed that I would participate in the event:
   - ○ >2 Weeks before
   - ○ 1-2 Weeks before
   - ○ <1 Week before

4. Were the event targets: Y N
   - Clear? ○ ○
   - Important? ○ ○

5. Were you prepared? Y N
   ○ ○

6. Did the team have everything it needed? Y N
   - Data ○ ○
   - Training ○ ○
   - Supplies ○ ○
   - Management support ○ ○
   - Other support (maintenance, IT, etc.) ○ ○

7. If there was anything missing, please describe:

8. Was your team successful? Y N
   ○ ○
   If "No," why?

9. Would you like to participate in another event?: Y N
   ○ ○
   If "No," why?

10. What would have made the event more effective?

Use other side if needed.

Figure 7-7. Kaizen event evaluation form. (For a blank form, see Appendix A.)

- Notification—a team member would not have had to miss an afternoon during the kaizen event because of child care requirements if he was notified of his participation sooner than one business day before the event. Better advance notice facilitates greater team member flexibility.

Relative to each of the identified "opportunities," the kaizen promotion officer should seek answers to the following questions:

- Did the abnormality (opportunity) occur because the kaizen event standard work did not address it?
- Did the abnormality occur because the standard work was not followed?
- Did the abnormality occur because the kaizen event standard work was not adequate?

Gemba Tales

A kaizen kaizen. Shortly after one of the first large-scale events at an explosive manufacturer, there was some shock and awe relative to the significant improvements and the rapidity and magnitude of the process changes. While everything was done safely and without any "incident" (industry euphemism), it made sense to ensure that the "management of the change" process, especially with regard to safety and quality, was integrated into the kaizen standard work. This was done in a kaizen with the appropriate stakeholders. While the output satisfied the mission, it also did wonders from a change management perspective, especially for those with regulatory compliance responsibilities.

To discern the answers, the kaizen promotion officer may need to review the "lesson" with its author(s) and ask for clarification and examples to identify the necessary countermeasures.

Sensei Feedback

A sensei's job, among others, is to teach, exhort, challenge, push, and cajole. The sensei will typically conduct some formal training on how to conduct a kaizen event, but invariably his primary focus is on the principles, systems, and tools of the lean business system; transformation of the culture; and results. The sensei expects the company's kaizen promotion officers to pay attention, ask questions, and learn how to plan, execute, and follow-through on their kaizen events.

A sensei conveys much of his teaching through verbal and, occasionally, written feedback. Often the feedback will encompass implicit, sometimes explicit, insights into what could be done to improve the company's kaizen event management performance. The feedback may seem personal, referencing things like leadership's lack of discipline and commitment, and the kaizen promotion officer's failure to drive adherence to kaizen event standard work. Nevertheless, this should not be taken discouragingly. The kaizen promotion officer who wants to learn how to be better at kaizen event management should ask, even hound, the sensei for instruction. Unfortunately, it is rare that a kaizen promotion officer or someone in leadership bluntly asks the sensei, "What did we do well?" and "What do we need to improve?" Direct questions will often yield direct and helpful answers.

COMMUNICATION

Effective communication in a lean transformation effort is absolutely critical. Good, frequent, consistent, multi-mode communication reinforces the change vision. It broadcasts the necessary early short-term wins that kaizen events accrue at the beginning of a lean launch. Good communication conveys how employees, through kaizen, have been empowered to take meaningful action and make meaningful change. This helps instill an invigorating culture of recognition and praise.

Communication is multi-faceted. There are at least two straightforward post-kaizen ways to share information with the local and the broader company audience: 1) the kaizen event summary report, and 2) the lean/kaizen newsletter. The summary report is a one- or two-page overview of the pertinent details of the kaizen. It can be easily displayed in a high-traffic area for others to scan. A periodic newsletter, distributed in hard or electronic copy, typically combines information relative to the broader lean transformation and summary results from kaizen activity around the facility, business unit, or corporation.

Kaizen Event Summary Report

The kaizen event summary report is a close cousin of the A3 report and the pre-event area profile. The intent is to quickly communicate the most basic kaizen information and give recognition to the team within one or at most two pages. Virtually every piece of information should be available within the kaizen report-out package, including the following summary report "staples":

- the team name/scope;
- team photo;
- list of team members and facilitator;
- kaizen targets with results;

Gemba Tales

"Letters" to leadership. A learned sensei often supplemented his means of feedback with handwritten letters. These letters were usually written in front of the kaizen promotion officer, no doubt as part of his training. The sensei then handed the letter to the kaizen promotion officer with the order to hand-deliver it to the (off-site) senior executive. The letters often summarized the kaizen activities, but were pointed when it came to the cultural barriers that the leader needed to address if he ever wanted a successful transformation. These barriers included concrete-head thinking like designing non-right-sized, ultra-complicated, batch-making, monolithic machines that were prone to extremely poor availability and performance, requiring cadres of technicians for continued life support.

- linkage to strategy deployment deliverables, value stream, or process improvement plan;
- key improvement ideas; and
- before/after photos or graphs.

As mentioned previously, the summary report can be displayed in a high-traffic area such as the break room, cafeteria, or the main entrance. It also can be easily transmitted via e-mail or posted on a company intranet.

Lean/Kaizen Newsletter

Monthly or quarterly newsletters are excellent communication, and thus change management, vehicles. Often the newsletter is prepared and distributed at the corporate, business unit, or facility level, providing that there is enough critical mass relative to size and lean activity. Modern publishing software can help the kaizen promotion office easily and cheaply produce a professional looking product.

The lean newsletter is an effective means for:

- communicating lean transformation vision, status, and plans as well as the linkage to company strategy;
- introducing new lean principles, systems, and tools;
- announcing new training offerings and resources; and
- recognizing kaizen teams (reflecting much that is contained within the kaizen event summary report) and individuals for jobs well done.

While the newsletter subject matter may be diverse, it is best to keep the entries brief and the overall length to two to four pages.

RECORD RETENTION

Kaizen embodies a "just do it now" spirit. Archiving, admittedly bureaucratic in nature, seems antithetical to lean. However, without some measure of archiving, it is difficult to:

- present powerful evidence of successful kaizens to the change management "antibodies";
- facilitate periodic value-added self-reflection (hansei);
- propagate or share the gains with other value streams, sites, valued partners, etc.; and
- if/when a kaizen is later conducted on the same target area (not uncommon in a lean enterprise), the archived documents can be referenced for pre-event planning purposes and to determine the magnitude of any backsliding from the prior event.

A company must first establish the standard work for kaizen archives. This encompasses, among other things, the:

- form/format, medium, and content of the material;
- the virtual and/or physical location of the archived material;
- the maximum lead time between the kaizen and submittal to the archives;
- the archive and archive process owner; and
- accessibility of the archives.

Care must be given to ensure that the archiving effort is extremely minimal. This means, for example, that no one should be engaged in the wasteful activity of recreating kaizen documents electronically after the event, unless it is a brief report such as the kaizen event summary report.

Kaizen Event Coding

At each of its sites, a lean enterprise will conduct numerous events throughout the months and years. To facilitate referencing, archiving, and retrieving past kaizen summary packages, it is useful to employ a simple kaizen coding system. The coding data format could be, for example, business unit, location, date, and kaizen team number. Accordingly, S&T-MV-8.3.YY-3 would indicate that within the systems & technologies business unit, at the Mission Viejo facility, a kaizen report-out was conducted on 8.3.YY, and it was the third kaizen conducted during that year. This data, plus a brief kaizen team scope/name should provide valuable insight for anyone accessing the kaizen archives.

SUMMARY

- Event follow-through is the fourth and final phase of kaizen event standard work.
- The post-event follow-through addresses issues of:

  1. sustainability,
  2. event management improvement,
  3. communication, and
  4. record retention.

- Sustainability is the most important of the four follow-through themes. For a kaizen event, sustainability is supported through the kaizen newspaper, leader standard work, and the post-event audit.
- There are three sources for kaizen event management improvement:

  1. kaizen event evaluations,
  2. lessons learned, and
  3. sensei feedback.

- Good, frequent communication reinforces the change vision and communicates how empowered employees, through kaizen, have taken meaningful actions toward substantive change. Communication about kaizen events is commonly done through a kaizen event summary report and/or a lean/kaizen newsletter.
- Record retention, while admittedly bureaucratic, is important because it provides powerful evidence of successful kaizen events, facilitates periodic value-added self-reflection, and helps share learning across value streams and locations.

PART III

DEVELOPING INTERNAL CAPABILITY: THE LEAN FUNCTION

8
THE KAIZEN-READY ENTERPRISE

"To become an able man in any profession, there are three things necessary—nature, study, and practice."
—Aristotle

MAKING KAIZEN YOUR OWN

A generic action plan for successful lean implementations has been outlined within lean literature (Womack and Jones 1996). The plan reflects a four-phased approach, the first of which exhorts the lean aspirant, among other things, to get lean knowledge. This is done initially largely by securing a sensei.

The second phase, "create a new organization," roughly prescribed as between the first six months of the lean launch and year two, calls for reorganizing by product or service family (becoming a value stream-based organization). Shortly after this reorganization, the plan requires the creation of a generically named "lean function." This function is none other than the kaizen promotion office (KPO). It goes by other names such as the just-in-time promotion office (JPO), lean office, company production or business system office, or continuous improvement office.

The lean function is a necessary resource for making an enterprise kaizen-ready. Readiness is largely achieved by developing a demonstrated excellence in kaizen event planning, execution, and follow-through. It is ultimately proven when the organization has transitioned into one that regularly employs principle-driven kaizen—events, projects, and daily kaizen. KPO establishment facilitates the transition from a sensei-dependent model to one that eventually will be employee driven. This is accompanied by dramatic changes in accountability and organizational development.

Certainly, the act of creating a lean function does not infer that the student has absorbed all that the sensei has to offer. Rather it recognizes that a student best learns how to ride a bike first by brief instruction and observation, then by practicing with the aid of training wheels and, eventually, by riding without training wheels. The removal of the training wheels in no way confirms expertise. Rather, it is a milestone that marks the beginning of a new lean maturity phase in which the sensei is used in a more strategic manner (see Figure 8-1).

Enablers

The transition to kaizen readiness is enabled by:

- sensei guidance,
- effective transformation leadership, and
- solid short-term wins.

The lattermost, a direct result of the first two elements and an important piece of successful change management, is critical for momentum while also serving as empirical evidence that internal kaizen resources are much more than self-funding. Kaizen is just good business.

The cost of sensei guidance should be understood as a long-term investment, granted that the intensity and leverage must shift over time. The sensei's experience and perspective can be invaluable when it comes to developing kaizen competency. As the organization moves forward on its lean journey, the sensei's role shifts from being predominately a kaizen event facilitator

Kaizen Event Fieldbook

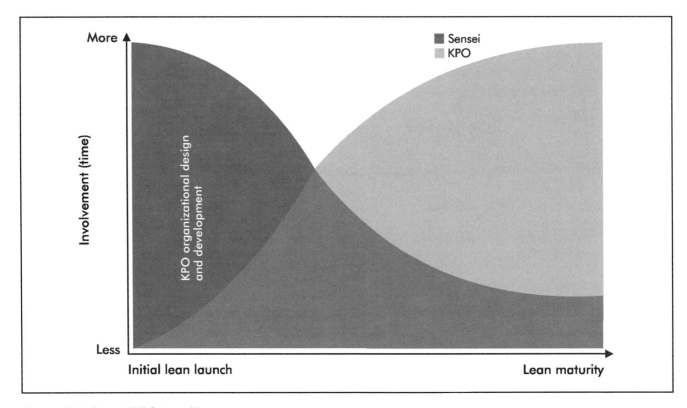

Figure 8-1. Sensei/KPO transition.

and trainer to a more substantial one of technical and transformation leadership coach. Coaching may be done in the "margins" of the kaizen event itself (which should get ever wider as the kaizen promotion officers undertake more of the facilitation) or within separate blocks of time. Coaching only provides real leverage if there is a strong internal team to absorb and apply it. It is often facilitated by sharing specific tools and techniques and assigning "homework" to be reviewed at a later date.

Effective transformation leadership is a prerequisite for kaizen readiness. Consistent with the lean implementation action plan, it entails establishing and nurturing a lean function. Of course, the lean function has meaning and relevancy only if leadership institutionalizes the rigor of a lean performance system and routinely applies kaizen standard work and strong change management.

Roadmap

Figure 8-2 reflects a recommended high-level roadmap for KPO development. Unfortunately, leaders often defer any meaningful application of the roadmap. While they may properly understand its importance, they frequently do not grasp the urgency. Why should they?

All may appear rosy as evidenced by an almost routine production of outstanding kaizen events and substantial sustained improvement in the targeted value streams. It is not that the results are not great; it is just that they could be better—better in terms of "ownership," capability development and scalability and, ultimately, cultural transformation. Unwittingly, leadership's lethargy, when it comes to developing the KPO, ends up retarding the enterprise's kaizen readiness. While the lack of desire to fix something that appears not to be broken is a barrier to roadmap progress, there are others. These barriers, real or otherwise, typically fall into the following categories.

- *Comfort*—the sensei-driven model, other than its costs, is relatively "easy."
- *Cost avoidance*—consulting costs are discretionary and can be cut relatively quickly,

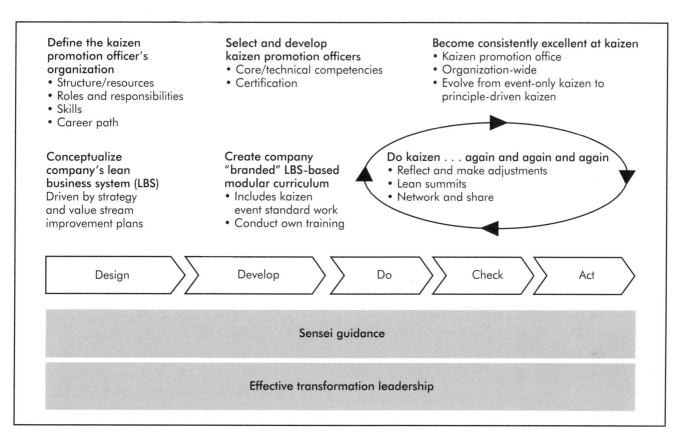

Figure 8-2. KPO development roadmap.

whereas headcount additions are much less so. Accompanying headcount are training and development costs, travel, etc. What often escapes many smart people is that, when deployed properly, returns on KPO investments are always favorable.

- *Executive A.D.D.*—attention deficit disorder among executives is a real, although not clinically documented, phenomena. Sometimes executives just cannot focus right now on developing the KPO. They will get to it . . . later.
- *Inexperience/ignorance*—not knowing how to attack the KPO development roadmap can facilitate procrastination.
- *People*—most permanent KPO resources, and certainly all of the temporary/rotational resources, should be provided from within the organization through redeployment. As waste is eliminated, work content is reduced, thereby freeing up employees to do other more value-added activities. The initial seeding of the KPO function is often difficult because the lean launch is fresh and the opportunity for redeployment may be slim. The kaizen promotion officers should be matched based upon key core and technical competencies. Unfortunately, the most qualified are usually busy in other roles and are difficult to redeploy. For strategic senior kaizen promotion officer roles, it often makes sense to bring in someone with extensive lean transformation experience from the outside (someone who has "been there, done that"). This represents a headcount addition—but one that is justified.
- *Power*—organizational structures, by their nature, assign and secure power. The KPO function is typically a new addendum to the current organization. As such, there may be some political maneuvering behind the determination of reporting relationships, headcount, and the like. Politics, no surprise, often slow down forward movement.

Lean Leaders' Short "Must Do, Cannot Fail" List

- Get a seasoned sensei to facilitate a successful lean transformation, *but* plan and act from day one to develop internal capabilities. Do not become sensei dependent!
- Staff a dedicated and decentralized KPO with high-potential employees. This should be done with virtually no net headcount additions.
- Invest in kaizen promotion officer development—time; coaching; hands-on, progressive experience; and formal and self-study.
- Support the KPO function by demanding organization-wide adherence to kaizen standard work and results, as well as expansion of daily kaizen activities.
- Use the KPO function as a "farm system" for lean leaders.
- Develop and "brand" your organization's own lean business system and curriculum as soon as possible.

Lean thinking seeks to optimize the flow of value in spite of the value-impeding functional barriers. Similarly, the KPO function should be situated within the organization so that it can best facilitate the flow of ideas and improvement.

KAIZEN PROMOTION OFFICE

Shortly after the first handful of kaizen events and, of course, the determination that lean is not a passing fancy, it is time to begin conceptualizing the KPO organization structure. This conceptualization may be "constrained" by the organization's current perspective of lean in general and kaizen specifically. In other words, if the lean effort is new and limited to a discrete value stream, location, department, etc., the perspective may be parochial—"That is something done in our Illinois operation . . ."

A mature lean view encompasses the entire enterprise and thus the entire organization. However, it is rare for a lean launch, especially within a large corporation, to rapidly propagate throughout the entire organization. Accordingly, pragmatism may dictate that the KPO organization structure germinate within a particular plant or office, knowing that as the lean effort expands to other parts of the enterprise, the KPO structure and roles will necessarily change. This is not to say that the organizational structure should be organic and reactionary. Instead, the KPO organizational design should be rational and done in a phased manner commensurate with the anticipated scope and path of the lean implementation. In any event, the KPO organization must be thoughtfully developed relative to structure, resources, roles and responsibilities, necessary competencies, and career paths.

Organizational Structure

The generic KPO organizational structure, reflected in Figure 8-3, should satisfy four basic criteria, without which the KPO function will end up as a fine example of organizational muda.

1. *KPO members should be dedicated.* While there may be a place for part-time kaizen coordinators who can help with many kaizen-related logistics and help facilitate certain daily kaizen activities, the real heavy lifting can only be accomplished by those who are full-time, dedicated KPO resources. Experience proves that a person cannot serve two masters well.

2. *Permanent resources should be limited, but sufficient.* Excess resources often introduce muda. Lean has no place for empire building and the related risk of "over-processing" when excess resources end up working on superficial stuff. On the other hand, a shortage of resources will hamper muda-eliminating activities and follow-through. One rule of thumb for KPO resource dedication is 1% to 2% of the total headcount. This figure is obviously not precise; it is subject to adjustment based upon value stream complexity and the size and required speed of gap closure, but it does give insight into the magnitude of the investment.

In situations where there is no critical mass (for example, a 40-person location), a dedicated, site-specific resource may not be pragmatic. In such a situation, it may

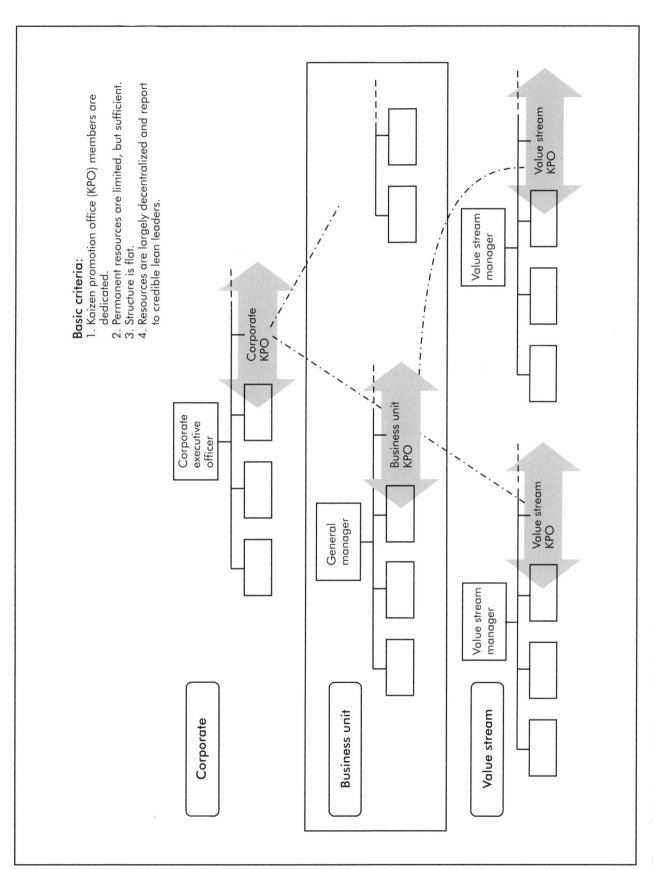

Figure 8-3. Generic KPO organizational structure.

>
> **Part-time usually means no time (or close to it).** Management understandably winces when it comes to new full-time assignments. Unfortunately, the "compromise" often ends up taking a high performer and adding part-time kaizen promotion officer responsibilities to an already full plate. What happens then is a constant battle between urgent and important. This dynamic was evidenced by a kaizen promotion officer who repeatedly had to remove himself from the midst of in-process kaizens. It seems that, in addition to being a kaizen promotion officer, he also was in charge of monitoring and purchasing critical raw materials, among other things. So, when the choice was to either shut down the plant due to a lack of raw materials or conduct a more effective kaizen, the former always won. Maybe they should have had a kaizen to implement a good lean material replenishment system so that the raw materials did not need babysitting by the kaizen promotion officer . . .

be appropriate to share regionally. For example, two relatively adjacent operations, when combined, may have enough critical mass and continuous improvement activity to fully employ a dedicated KPO resource. While this may require a level of KPO travel, this is neither abnormal nor unwarranted.

Some organizations, in the absence of site-specific critical mass, will deploy a kaizen coordinator. The coordinator, typically part-time (yes, it violates the first criterion!), is a type of junior kaizen promotion officer whose activity is often limited to kaizen event logistics, some pre-planning and follow-through elements, and the facilitation of daily kaizen activities such as kaizen circles. This position can serve as a relatively low-risk proving ground for future full-time kaizen promotion officers. Coordinator leverage, no surprise, is limited or enhanced by the level of commitment from local leadership. The coordinator also should be supplemented by a full-time kaizen promotion officer, whether on- or off-site.

The notion of permanent resources does not extend to those people who may serve temporarily as members of a KPO pool. This pool is comprised of employees who have been redeployed from their regular positions due to productivity improvements. They are expected to conduct and support kaizen activities only until they are permanently reassigned to other positions. If the organization develops virtually everyone as lean generalists as it should, the kaizen promotion officers are catalysts and facilitators within what should eventually be a lean culture.

3. *KPO organizational structures should be flat.* Hierarchy equals bureaucracy. While a lean transformation requires a corporate kaizen promotion officer to ensure standardization across the enterprise, serve as the internal head sensei, work change management issues, etc., the KPO structure should be comprised almost exclusively of "doers." As such, most of the KPO resources should be focused within the value stream. Corporate- and business-unit level positions are often single contributors.

4. *KPO resources should be largely decentralized and report to credible lean leaders.* The KPO is for high-leverage action, not ornamentation. Value stream-based organizations routinely outperform those that are centralized and highly departmentalized; the KPO structure is no different. KPO resources should report (solid lined) to business-unit general managers and value stream managers. These resources work on strategy deployment deliverables and relevant improvement plan activities. It is appropriate to maintain a dotted-line relationship back to the corporate/business-unit KPO for the purpose of standardization and alignment in methods and curriculum, technical development, sharing of best practices, and the like.

Roles and Responsibilities

The kaizen promotion officer role and responsibilities can be simply articulated within the context of seven key result areas. These areas represent the critical "outputs" of the generic kaizen promotion officer position, which are driven by specific, measurable, actionable, relevant, and time-bounded (SMART) deliverables. Understandably, the underlying activities, emphasis, and responsibilities vary based upon the particular kaizen promotion officer's position within the organization (corporate, business unit, value stream, etc.) as well as the specifics of the lean transformation status and related challenges. The following provides some insight into the kaizen promotion officer's role and responsibilities (see Figure 8-4).

1. *Change management.* By definition, the kaizen promotion officer is in the business of change management as an advisor, trainer, coach, and catalyst. Indeed, *kaizen means change.* This change is effected through formal and informal training and education, facilitation of the kaizen event process and daily kaizen activities, written and oral communication, and the deployment of alignment mechanisms (for example, strategy deployment, lean management systems, and steering committees) that drive focus, accountability, and sustainability. Finally, the kaizen promotion officer is called to a professional life of constant exhortation and spreading of the lean gospel. An example change management deliverable: *Facilitate implementation of pilot lean management system by 5/30/YY.*

2. *Business system and curriculum development.* The context for kaizen is the organization's lean business system. The KPO function comprises the enterprise's team of dedicated lean technical experts. Accordingly, shortly after the lean launch, the company, as facilitated by the kaizen promotion officer, should define its business system, the underlying alignment mechanisms,

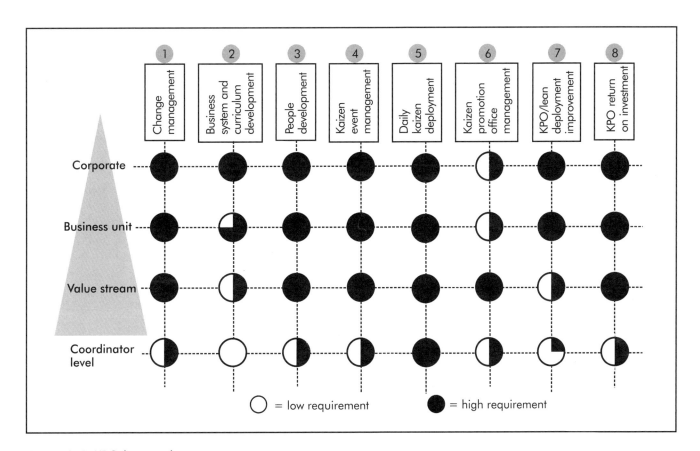

Figure 8-4. KPO key result areas.

specific tools (as "pulled" by its strategic imperatives and value stream improvement plan[s]), and the system's overall balance and orientation. See Figure 8-5.

There is little reason to reinvent the wheel. Along with the rework associated with reinvention, there is significant risk of unwittingly marginalizing or even gutting the heart of the Toyota Production System. Nevertheless, business system development serves two purposes: 1) it is a profound learning exercise, and 2) change management, enabling the enterprise to better communicate the system in relevant terms and examples.

Once the business system is developed, the kaizen promotion officer builds out the various business system element training modules. These modules typically include: 5S (sort, straighten, shine, standardize, sustain), visual controls, standard work, setup reduction, kanban, value stream mapping, kaizen event standard work, problem solving, etc. Often modules can pre-exist the business system model, basically because they are "no brainers." The need for other modules will be manifested over time. For example, every lean enterprise needs 5S rather immediately, while 3P (production preparation process) training is often not required until later. Any sensei worth his salt will have module materials that can be borrowed and re-tooled by the kaizen promotion officer or jointly with the sensei.

Sometimes organizations distribute the module development effort among many by employing a lean champion strategy. Under this strategy, individuals, not necessarily kaizen promotion officers, are selected based on their motivated ability to develop a specific

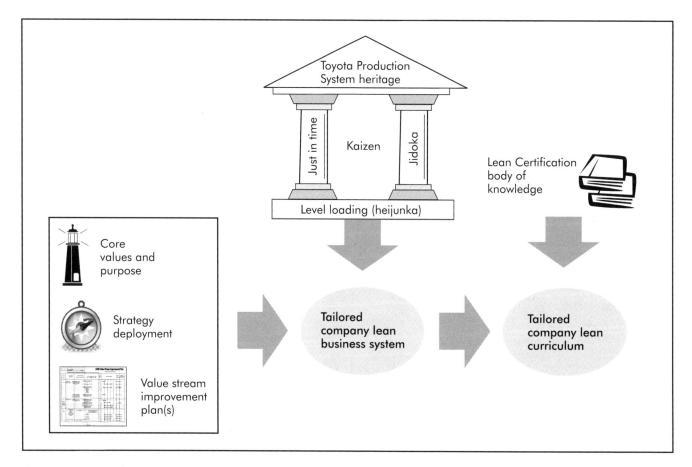

Figure 8-5. Lean business system and curriculum development.

company-tailored lean tool training module. They author and maintain the module and its materials, create recommended reading lists, kick-off the training, and train other trainers. This approach helps to distribute "ownership," thereby expanding the lean culture. Example deliverable: *Develop/deliver heijunka module by 7/15/YY.*

3. *People development.* To transition from a sensei to employee-driven kaizen enterprise, people must be developed at both an individual and corporate level. The kaizen promotion officer serves a critical role in the lean technical and behavioral skill development of employees at various strata within the organization and at different times and venues.

Corporate, business unit, and value stream management team levels have functional peers (sales and marketing, finance and administration, operations, engineering, human resources, etc.) with whom the kaizen promotion officers interact on a frequent basis. The development opportunities are presented within day-to-day (for example, staff meeting and/or daily accountability process and daily kaizen activities) and breakthrough management activities (for example, strategy deployment checkpoints and kaizen events) as well as during formal training.

Further, kaizen promotion officers, as internal sensei, are obligated to mentor others at a more intensive level. Pragmatically, these efforts must be focused and typically are limited to other kaizen promotion officers, kaizen promotion officer candidates, and others who have a demonstrated need for deep lean knowledge. The SME/AME/Shingo Prize Lean Certification requires mentoring as part of the criteria for the silver and gold levels. Example deliverable: *100% of value stream members trained in Lean 101 by 9/30/YY.*

4. *Kaizen event management.* The kaizen promotion officer is the subject matter expert, guardian, and facilitator of the standard work that constitutes the multi-phase kaizen event approach. This standard work should be appropriately tailored and applied within the particular enterprise and its value streams. The senior-level kaizen promotion officer should serve as the liaison with the external sensei.

Gemba Tales

Champion and reality show star. A maintenance engineer was "volunteered" to be the setup reduction champion. He developed an outstanding training module, complete with a video to demonstrate the situation before and after the setup reduction activity. He wisely decided to video a process that virtually everyone could relate to—changing the tire on a car. Short on available actors, he used himself as the "operator." The video, to say the least, was insightful. He demonstrated, among other things, the benefits of point-of-use storage, proper tool selection, mistake-proofing, etc. However, he definitely gave more than was expected . . . module trainees will never forget the champion's "plumber-type" exposure of his backside when he was bending over to change the tire.

Example deliverable: *Leader standard work developed/updated as part of all kaizen event follow-through phases, starting 3/31/YY.*

5. *Daily kaizen deployment.* To achieve an advanced level of lean transformation effectiveness, an enterprise must evolve from system-driven kaizen to principle-driven kaizen. This requires a transition from a purely kaizen-only event model to a combination of kaizen events, conducted in accordance with standard work, and daily kaizen activities, projects, and "just-do-its." The daily kaizen activities (see Appendix B) encompass mini-kaizen events, kaizen circle activities, suggestion systems, 5S activities, etc. The related daily kaizen opportunities are identified by means of an effective lean management system and through the eyes of engaged, empowered, and trained employees as they constantly compare the current state with an envisioned ideal state. The KPO's role is to help: 1) facilitate the adoption and maintenance of the lean management system (including the suggestion system), 2) assist in the training and development of problem-solving employees, 3) facilitate activities such as mini-kaizens and kaizen circles, and 4) train others within the organization so that they may train and facilitate and thus propagate a daily kaizen culture. Example deliverable:

Develop and pilot kaizen circle activity standard work/training by 7/15/YY.

6. *Kaizen office management.* The kaizen office is both the physical and virtual space in which the kaizen promotion officer operates; it is predominately virtual because that easily accommodates the various gemba locations. Office management extends to the dotted-line reporting relationships between the different decentralized kaizen promotion office levels. Logically, the focus of office management is primarily on the key result areas, the related issues (change management challenges), and opportunities (for example, propagating best practices across the company). As with any staff position, special care should be taken to avoid the muda of bureaucracy, especially that of over-reporting.

Due to the purposeful "leanness" of the kaizen promotion office, direct reporting relationships are rare and usually limited to management of the temporary kaizen pool resources. Kaizen pool management often includes training, project assignment, supervision, and basic human resources administration—schedule rotation, prevailing pay rates, etc.

Office management also includes maintenance of the various kaizen supplies, training materials, books, DVDs, filing of physical and electronic kaizen results, performance tracking, and development and issuance of the various newsletters. The office will often have responsibility for a "moonshine shop." This shop is an on-site laboratory for product and process development (think of it as a MacGyver-type place to tinker). It is usually squirreled away somewhere on the plant floor or office. There creative, multi-skilled folks can generate low-cost, quick turnaround prototypes out of materials like wood, PVC pipe, cardboard, foam, coated tubular steel, and contributions from the "boneyard" (discarded things like equipment, tooling, instruments, and electronics) using basic hand and power tools or low-volume, flexible equipment such as a lathe. The administrative or office equivalent of the moonshine shop provides the ability to reconfigure cubicles and storage units. It also includes an information technology capability to trial, often in a test environment, software code modifications and/or develop spreadsheet and spreadsheet macros to conceptualize more permanent future-state fixes, etc. Example deliverable: *Develop KPO pool pay and rotation policy by 8/31/YY.*

7. *Kaizen promotion office/lean deployment improvement.* This key result area is largely about "kaizening" the kaizen process and sharing best practices throughout the enterprise. The former is essentially event management and daily kaizen improvement extended to address opportunities throughout the enterprise. The forums for such collaboration and improvement can include quarterly "Lean Summits" in which the kaizen promotion office members meet for several days and work on specific high-leverage items, often supplemented by pre-work. Best practices relative to process and product improvements can be shared in similar ways as well as via periodic teleconferences, web-based meetings, and other electronic means. Example deliverable: *Establish quarterly kaizen promotion office Lean Summit by Q3.*

8. *Kaizen promotion office return on investment (ROI).* While the first seven items may be dismissed as overly qualitative in nature, this one is quantitative. In fact, if done correctly, satis-

---

Gemba Tales

How do you define "KPO?"

Management quickly grasped the transformation opportunities offered by lean after the second kaizen event. Zeal was reflected in management's desire to immediately apply some not insignificant resources to a newly established lean function (the kaizen promotion office). The KPO was quickly staffed with four high performers. However, in their haste there was a fair amount lost in translation, not the least of which was the full role and nature of the kaizen promotion officers. This was evidenced as one of the four was introduced at a kaizen event as a "KPO." When a participant innocently asked what that was, the kaizen promotion officer was unable to decode the three-letter acronym . . .

faction of the first six should drive satisfaction of kaizen promotion office ROI. An example deliverable: *Kaizen promotion office ROI for the TLX value stream > 5× for 20YY.*

Competency

With the KPO key result areas or "outputs" established, it makes sense to next determine what is necessary to satisfy them. Certainly, there are a number of contributors to success, many directly driven by effective leadership, including:

- sufficient resources,
- role clarity within the organization,
- culture—sense of urgency, rigor, accountability, and
- focus.

Presuming that leadership is doing the right things and doing them well, the most important driver of key result area satisfaction is getting and/or developing kaizen promotion officers with the requisite core and technical competencies. See Figure 8-6.

Many people, if asked to list the necessary kaizen promotion officer skill sets would instinctively and immediately launch into the lean technical realm—tools and techniques . . . with lots of Japanese names. This instinct is misguided. Effective kaizen promotion officers are built upon a foundation of strong core competencies. Provided that the candidate has that foundation as well as technical aptitude and the motivation, the rest can be learned. Technical lean is *not* rocket science; it just requires a tremendous amount of study and observation and, most importantly, learning by doing.

Core Competencies

Core competencies represent a suite of performance skills essential, in varying degrees, for a broad range of jobs. They largely reflect work habits, attitudes, behaviors, and personal characteristics. While many core competencies are acquired early in life, they can be further developed and honed. They reflect really "how" a person performs a particular job and may be contrasted with technical competencies, which define the "what" or content of the job in terms of knowledge, tools, systems, instruments, and equipment.

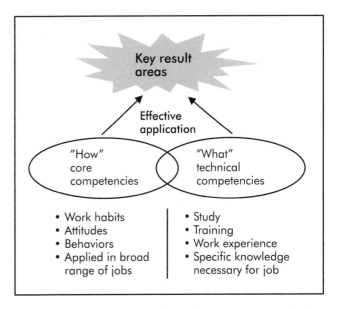

Figure 8-6. Core and technical competencies for the key result areas.

There are a large number of core competency categories with numerous variants, some substantive, some driven more by nuance or semantics, and some representing subsets or super-sets of a particular category. In other words, there can be an exhaustive number of permutations. Accordingly, the reader should not get too wrapped up in the names used, but rather focus more on the substance and underlying themes.

Figure 8-7 lists the eight required KPO core competency levels by placement and role within the organization. Note that the levels are more relative than absolute. Consistent with the notion of "core," they should be shared across a number of roles within any enterprise engaged in a credible lean transformation. Pragmatically, not every role requires the same level of proficiency or strength for every core competency. This is true even within the different levels of the kaizen promotion office.

1. *Strategic orientation.* Kaizen is a blind faculty in that it must be directed at a particular scope and opportunity for specific purposes. Thus, at a formal level, this is where strategy deployment and value stream improvement planning are engaged. The kaizen promotion officer must promote and maintain a long-term, relevantly broad-based

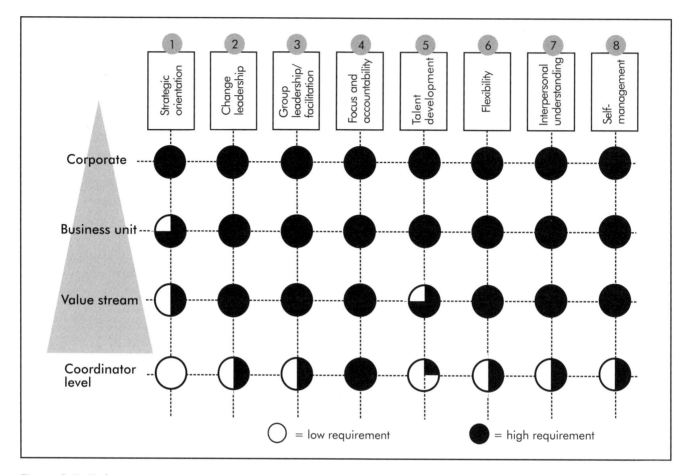

Figure 8-7. Eight core competencies.

perspective on the business and value stream to effect high-leverage improvement. The KPO should necessarily have a developed level of customer focus and insight.

2. *Change leadership.* This is likely the most important competency of the kaizen promotion officer. It is also the most daunting, given that virtually everyone in the organization does *not* report to the kaizen promotion officer. This requires the ability to alert and energize individuals and groups to the ongoing need for change, while at the same time persuading and influencing them to learn and deploy the principles, systems, and tools of the lean business system to identify and achieve specific change.

Change leadership requires an excellent grasp and application of the change process itself (see John P. Kotter's writings for some best practices), along with:

- Spoken communication—this is the ability to clearly, accurately, and candidly present information and influence or persuade others through oral presentation, whether the circumstances are positive or negative. This skill encompasses the ability to listen well and is rooted, like lean, in humility and respect for humanity.
- Assertiveness—the kaizen promotion officer must be able to maturely express his feelings and opinions in spite of disagreement. Further, he must be able to accurately communicate to others regardless of their status or position. This occasionally extends to the need to be uncompromising, especially when it comes to certain non-negotiable elements of lean.
- Reading the system—this is the ability to recognize and use information about organizational climate and key individuals

to accomplish legitimate organizational goals. The kaizen promotion officer must be cognizant of the importance of timing, politics, and group processes in managing change.

3. *Group leadership/facilitation*. The kaizen event process and many daily kaizen activity approaches require both leadership and facilitation. The kaizen promotion officer needs to know the difference between the two and when best to use one versus the other. A facilitator serves as the process leader and manager of group dynamics so that the content experts can more effectively achieve their objectives. On the other hand, leaders, while directing the process (more or less) in the absence of a facilitator, own the content and seek to lead the group to achieve the team's objectives. It is worthwhile to provide formal facilitator training to all kaizen promotion officers.

4. *Focus and accountability*. This competency should be applied by the kaizen promotion officer to drive performance spanning two levels: break-through and day-to-day. The breakthrough encompasses the achievement of the key result areas and related deliverables, the execution of relevant strategy deployment deliverables, and the achievement of value stream improvement plan objectives. Day-to-day performance includes that related to kaizen event execution, sustainability of prior improvements, deployment of daily kaizen activities, and execution of various tasks.

   Success requires the ability to start and persist with a specific course of action, embodying motivation and a sense of urgency. A good kaizen promotion officer will have a "whatever it takes attitude" with a willingness to commit to long work hours and personal sacrifice.

5. *Talent development*. The kaizen promotion officer is a genuine teacher who must foster the learning and long-term development of others. The "student body" extends to leadership, peers, subordinates, other kaizen promotion officers, kaizen team participants, all employees, valued suppliers, and the broader community. Virtually every time the kaizen promotion officer has contact with these people there are teachable moments; some are formal and intensive, others informal, simple, and brief.

   In any event, the teaching must be orthodox, consistent and, when appropriate, it must demand accountability relative to application and ownership. The SME/AME/Shingo Prize Lean Certification recognizes the value of development. It requires candidates, among other things, to mentor one or more persons to earn the silver and gold certification levels. According to the mentoring guide (SME/AME/Shingo Prize Lean Certification Mentoring Guide 2006), "[t]he primary objective of mentoring is to enhance the junior party's knowledge and abilities so that he is better able to perform a function, in this case, to deliver results using the tools and techniques of lean thinking."

6. *Flexibility*. An effective kaizen promotion officer will adapt and work effectively within a variety of situations. These situations often present unique lean application opportunities, at least relative to the kaizen promotion officer's experience portfolio. The most important thing to remember is that the lean principles always work. With the proper measure of creativity, sensei input, and adherence to the kaizen system, the lean countermeasure will yield improvement.

   While technical challenges require flexibility, change management challenges often require simply "coping." The kaizen promotion officer will routinely have to maintain a mature problem-solving attitude when dealing with interpersonal conflict, personal rejection, hostility, and time demands. Of course, flexibility does have its limits.

7. *Interpersonal understanding*. The kaizen promotion officer must have a developed awareness of the needs and views of others. This requires a level of emotional intelligence, especially in situations in which kaizen team members and other employees are anticipating or experiencing substantial change.

8. *Self-management*. While the kaizen promotion officer is a teacher, she is also a student. The kaizen promotion officer is

responsible for continued self-growth and development while also maintaining an awareness of the impact on others. Lean transformation activities will present the kaizen promotion officer with a constant supply of diverse learning opportunities.

Of course, before delving into the specific competencies, the invariable question must be addressed, "How can these 'soft' competencies be measured?" The "Harvey balls" (see Figures 8-4 and 8-7) give some insight relative to high, medium, or low requirements, but that is a subjective realm. Because of the "core-ness" of the competencies, the enterprise should endeavor to define them and provide a reference to assist people in determining job-specific requirements and assessing a candidate's competency level for all relevant jobs (not just KPO jobs). Figure 8-8 provides a core competency evaluation reference.

Technical Competencies

The technical competencies for lean are more concrete than the "softer" core competencies; however, they are still somewhat elusive. Two obvious and important questions are: 1) "What material should be learned?" and 2) "How can understanding and proficiency be measured or validated?"

The best teacher is experience (learning by doing). However, this alone is not a sufficient or at least an efficient means for developing technical competencies. There needs to be a balance between knowledge, application, and coaching. The combination of these three, set upon a foundation of core competencies, develops true technical competency.

Absent a dedicated, personal sensei, one of the best ways to develop lean technical competency is to pursue the SME/AME/Shingo Prize Lean Certification and supplement it with other study and experience. See Figure 8-9 for an example of how these elements may be combined to provide a structure for developing kaizen promotion officer technical competency, while at the same time answering the two previous questions relative to what to learn and how to validate the learning.

The (cumulative) multi-level Lean Certification combines:

1. knowledge requirements as outlined in the body of knowledge (summarized in Figure 8-10),
2. various project work with increasing leverage, and
3. mentorship at the silver and gold levels.

Access the SME website (www.sme.org/certification) for further insight and information, including:

- recommended reading lists,
- portfolio requirements,
- portfolio samples,
- body of knowledge, and
- mentoring guide.

While the Lean Certification "curriculum" is substantial, it may not be sufficient for the specific needs of the KPO organization. The knowledge-based exams will not necessarily drive the requisite "depth" of knowledge in the areas most pertinent

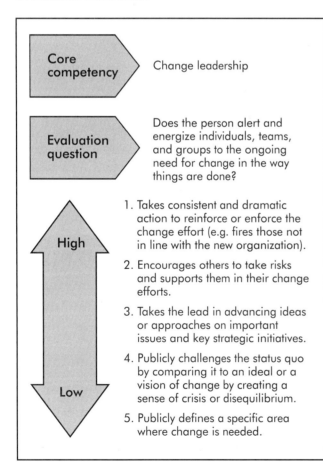

Figure 8-8. Example core competency evaluation reference.

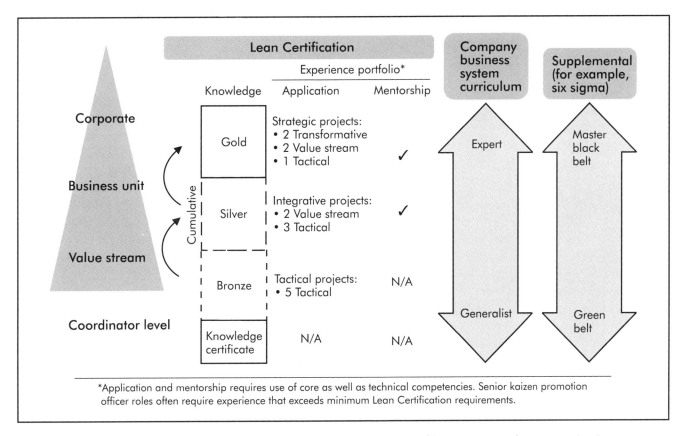

Figure 8-9. Example kaizen promotion officer technical competency profile in a mature lean organization.

to the enterprise's value stream(s). Accordingly, the kaizen promotion officers should know, and be able to teach, the company's tailor-made lean business system curriculum. Further, kaizen promotion officers, depending upon their level and career path, should supplement their lean knowledge with other synergistic knowledge. An obvious candidate topic is six sigma. Others include project management and facilitation training.

Other Considerations

In addition to core and technical competencies, there are other "job description" elements that must be considered. These include the demands of the kaizen promotion officer's job, prior experience, and education requirements.

Simply put, the kaizen promotion officer's job is demanding. It requires long hours, especially during kaizen events, frequent travel (if the responsibility spans multiple locations), exposure to sometimes less than comfortable physical environments (wherever the gemba is), and the stress of driving change, often with people who do not want change. That said, most kaizen promotion officers will say that, despite those challenges, they love what they do!

Job experience in the kaizen promotion officer's role is indispensable and necessary to meet its specific challenges. The Lean Certification criteria require a combination of experience in industry and lean implementation as well as time in the study of lean. The *combination* of these experience requirements are 4, 6, and 8 years, for bronze, silver, and gold, respectively. As evidenced by the following sampling of some kaizen promotion officer job description extracts, the theme of prior experience, as well as specific education and certifications, is important.

Sample kaizen promotion officer job description criteria extracts:

- BS or higher in technical discipline of mathematics or engineering;

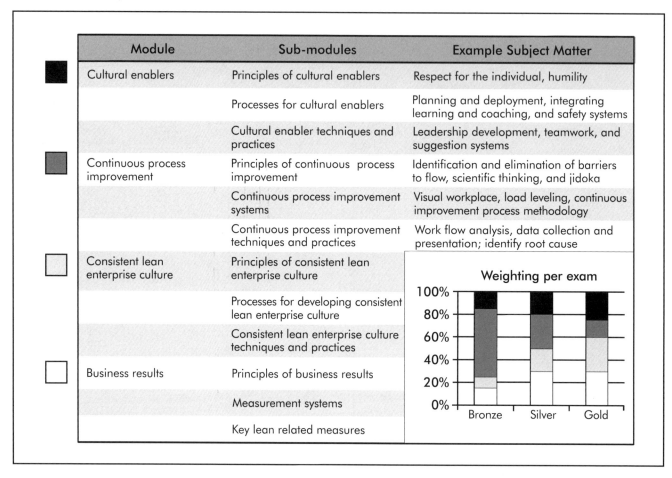

Figure 8-10. Technical competencies: Lean Certification body of knowledge.

- six sigma green belt or black belt certified;
- BS in engineering or related technical discipline;
- work in the Toyota Production System (TPS), Danaher Business System (DBS), or other recognized lean business system;
- broad leadership roles from disciplines such as engineering, quality, or product management;
- demonstrated mastery of TPS tools (5S, TPM, kanban, kaizen, value stream management, visual factory, poka-yoke, heijunka, single-minute exchange of dies [SMED], etc.)

Career Path

As previously outlined, the KPO organization should be flat, lean, and decentralized whenever and wherever possible. Accordingly, the kaizen promotion officer's career path, for all but a few who will stay within the discipline, eventually leads outside of the promotion office. This, especially from the perspective of the lean leader, is a wonderful thing for the lean transformation. After a minimum of roughly 18 to 24 months, a kaizen promotion officer within any lean-worthy enterprise and with a substantive experience portfolio, will have gained an incredible education, all while helping drive substantial improvements within the company.

When a kaizen promotion officer is reassigned outside of the promotion office, it is typically to a more senior role, one to which he will take his now expanded competencies and perspective, and leverage them within a line position or some other critical support role. Many senior executives rightly see the kaizen promotion officer function as a developmental path for high-potential employees.

KAIZEN PROMOTION OFFICER SELECTION

Kaizen promotion officer selection is extremely critical. It is a high-impact position and one for which, due to the leanness of the KPO organizational design, there are few resources to cover any weaknesses. After having defined the kaizen promotion officer roles and responsibilities, and the required core and technical competencies, the next, most rational step is to apply them to the selection process. Depending upon the size of the enterprise's performance gap, the desired speed for gap closure, and the culture's ability to absorb outsiders, the organization has to decide whether to develop/find its kaizen promotion officer resources from the inside, outside, or a combination of both.

Before embarking on a formal interview and evaluation process, the organization should not overlook the best internal candidate identification and assessment tool there is—kaizen event performance. The kaizen event represents a custom-made laboratory for secret auditions. Within the span of a kaizen an educated observer can quickly separate the wheat from the chaff. Worthy kaizen promotion officer candidates naturally respond to the kaizen process as if it is hard-coded within their DNA. They almost intuitively grasp the tools and techniques; they eagerly wage war on the waste they identify, apply "MacGyveresque" creativity, and revel in the opportunity to make dramatic improvements within a short few days. Sure, most of the time they are pure neophytes to lean in general and kaizen specifically; but they "get it" and the rest, provided that they are a substantial match with the core competencies and have the technical aptitude and desire to develop.

The same assessment strategy, although obviously a lot less covert, can be applied to evaluate outside candidates. Bring them in for an extended interview for a day or two during a kaizen event. The company can have the candidate deliver pre-event training and co-facilitate a portion of the event. This may seem like an unorthodox interviewing method, but unfortunately there are many people who mistakenly proclaim themselves as lean experts. This false advertising is not necessarily purposeful in nature. It is more likely a deficit of self-awareness of their knowledge and experience gaps, often due to their limited lean application in a single industry. In other words, they "don't get out much."

Any worthy kaizen promotion officer candidate seeking a position within an organization will require an extensive trip to the gemba(s). The candidate should also aggressively probe the sufficiency of the prospective employer's lean leadership and talk to employees at all levels of the organization to discern if the company "walks the walk." If the candidate lacks this rigor and intellectual curiosity, then he is neither experienced nor savvy and should not be considered for anything but a junior kaizen promotion officer role.

Regardless of whether the positions are filled internally or externally, kaizen promotion officers require continuous development. They are principally knowledge workers and to effectively coach those within the organization, their knowledge must be (largely) ahead of everyone else's. While the senior kaizen promotion officers should shoulder some responsibility for developing the junior kaizen promotion officers, the primary responsibility for development lies squarely with each individual. Consider it the eternal pursuit of self-kaizen. This personal development curriculum necessarily encompasses core and technical

> **Gemba Tales**
>
> **Learning requires work.** We were doing a setup reduction on a 500-ton injection molding press. It was Thursday morning and the team had reduced the setup from 2.5 hours to 17 minutes. The sensei held up his hand showing five fingers. I asked the interpreter, "What does this mean?" The interpreter related that the team needed to get to 5 minutes. I passed this on to the team members and told them to try again. By lunch time, the team had reduced the setup to 10.5 minutes. Again, the sensei held up five fingers. I asked him, "What other things should we look at?"
>
> The sensei replied, "Craigsan, if I give you answers, what will you learn?" To say the least we went back and looked harder. Ultimately, we reduced the setup to 4 minutes and 23 seconds. Too often we want the answer given to us without doing the work necessary to truly learn and understand.
>
> *Tale shared by Craig Robbins*

competencies and should be multi-faceted. The intellectually motivated kaizen promotion officer will diligently and creatively seek learning opportunities from a number of sources and experiences, including:

- Practicing application—there is no better way to learn, whether it may be by facilitating kaizen events or daily kaizen activities, developing training material, or delivering workshops. Failure often teaches more profoundly than success.
- Study—the lean library continues to expand dramatically. There are many excellent books that the kaizen promotion officer should read and, in some instances, re-read. The Lean Certification path requires a rigor of study that is typically missing from the normal casual self-study. Additionally, there are numerous workshops, webinars, periodicals, and web articles offered by organizations such as the Society of Manufacturing Engineers (SME), Association for Manufacturing Excellence (AME), Shingo Prize, Lean Enterprise Institute, American Production and Inventory Control Society (APICS), and American Society for Quality (ASQ).
- Observation—direct observation is a simple, but powerful learning method. Go and observe others as they teach and facilitate.
- Mentorship—the responsibility of mentorship sharpens the mind by, among other things, forcing a person to think about the way to best assist others in their developmental journey. In turn, this often prompts self-reflection and growth. On the flipside, it is a great opportunity and privilege to find a qualified mentor. A good mentor is a great treasure.
- Networking—the lean community is filled with individuals who, having learned from others, seek to return the favor and share their knowledge. The opportunities are far and wide, formal and informal, extending to web logs, seminars and site visits, professional meetings, phone calls and emails with friends, colleagues, acquaintances, and friends of friends. The kaizen promotion officer's network should not be parochial. In other words, it is just as important to network outside of your industry as it is inside. Lean principles transcend industry. By studying practices in disparate value streams, technologies, products, services, cultures, etc., the observer often gains new and deep insight.
- Benchmarking—site visits, within plant, office, lab, or field, provide a unique opportunity to compare lean applications. While most visits include aspects that are less than the lean gold standard, there is virtually always something that the visitor can take back to make his company better. A word of caution, however, is in order: Guard against falling into a trap of casual industrial tourism to ensure that each benchmark visit is conducted with an agenda and appropriate rigor.
- Volunteerism—there are opportunities for the kaizen promotion officer to freely share his time and simultaneously further develop his knowledge. Examples include serving as a Shingo Prize examiner, assisting in the SME/AME/Shingo Prize Lean Certification process, etc.
- Being a "sponge"—this is a euphemism for drinking in all that you can. It includes going on gemba walks with experienced lean thinkers and following at the elbow of a visiting sensei. Where there is a will to learn, there is a way.

Clearly, the opportunities and modes for learning are manifold, but what about specific preparation to become a kaizen event facilitator? In addition to some basic study, it is best learned through assuming a progression of roles and responsibilities over a number of kaizen events (see Figure 8-11). During this progression, it is important to experience a variety of kaizen event scopes and types. For example, a path that begins with five successive setup reduction kaizen events does not provide the requisite amount of variety and learning opportunities for the facilitator in training. Developing a multi-dimensional facilitator requires exposure to and participation in a broad mix of kaizen events: flow, process, product, service, transactional, etc.

While most facilitators in training are understandably impatient to get on with it, their immersion should start with at least one or two

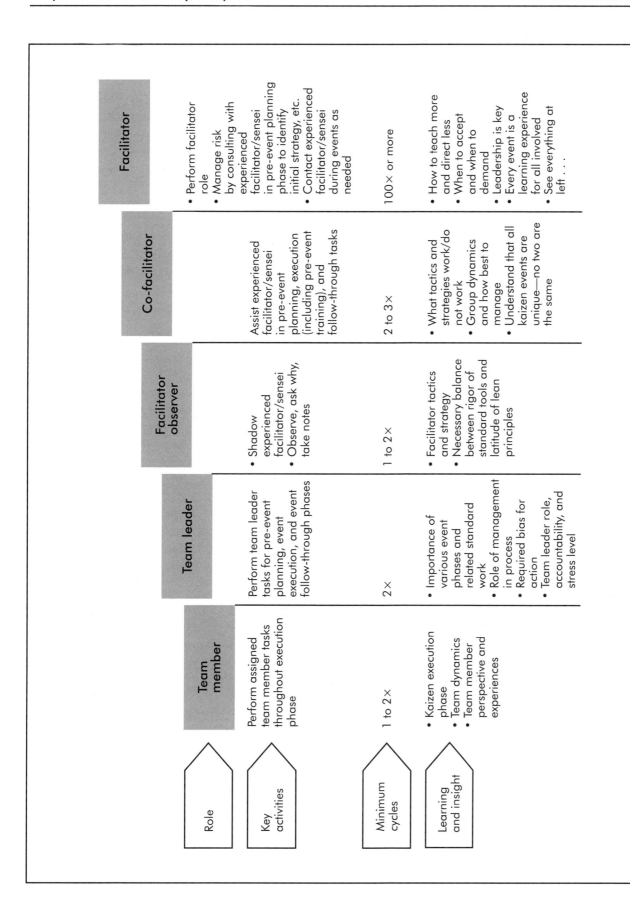

Figure 8-11. Example kaizen facilitator training path.

cycles as a kaizen team member. How better to apply the basic kaizen tools and gain the perspective of the kaizen team member? The next role to experience is that of the team leader, preferably for two events or more. The team leader's scope of responsibility is broader than the team member's, encompassing almost the entire multi-phase kaizen standard work and experiencing a substantial change in accountability and stress. Much of a facilitator's focus and coaching is directed at the team leader during the event (it is the highest leverage way to influence team strategy and behaviors). Accordingly, it is critical for the facilitator in training to experience the team leader role firsthand.

It is only after experiencing the first few cycles that the facilitator in training should begin the direct study of event facilitation, which follows the progression of observer, then co-facilitator and, ultimately, facilitator.

The facilitator observer step includes direct observation of an experienced facilitator/sensei during the entire event execution phase. This should be done at least one to two times during which the observer should take notes and discreetly ask "why?" many times. The primary risk in the facilitator observer step is boredom that begets a lack of attentiveness and thus missed learning opportunities. Yes, observing a kaizen is sometimes like watching grass grow or paint dry, but in between the sensei interaction with the team and management, there is substantial time to "pick the brain" of the sensei. It is during this time that the student should be gaining insight into training methods, facilitator tactics, and strategy, seeing firsthand how the sensei balances the rigor of standard tools against the latitude of lean principles.

The co-facilitator step, which should be repeated at least two to three times, requires a much greater level of involvement and responsibility. The co-facilitator's scope extends beyond the execution phase to include pre-event planning and follow-through. It also requires the co-facilitator to conduct a portion of the pre-event training. Anticipation of the trainer role should induce some level of serious study. Stress can be good.

During the co-facilitation step, the facilitator in training should start to gain insight into the tactics and strategies that work (or do not) and how group dynamics are managed. There also is a realization that no kaizen event is exactly the same. This is usually accompanied by, "Hey, my notes say that at this point I am supposed to . . . !?!" The cookbook mentality is a dangerous thing and the sooner this is learned, the better.

The facilitator step is the point at which the training wheels come off. The real and obvious risks can be mitigated early on by thorough pre-planning and consultation (sometimes real time—the cell phone can be a wonderful tool) with an experienced facilitator/sensei throughout all phases of the kaizen event. The new kaizen promotion officer should not expect pre-packaged answers. A good sensei will make use of the Socratic Method.

As mentioned in Chapter 1, many believe that before a facilitator approaches a basic level of expertise, he needs to have coached a minimum of 100 kaizen events. Even if this amount were cut to 50, depending upon the density of the kaizen schedule, it would easily take up to 2 years. To accelerate and broaden the learning experience, it may make sense to have the inexperienced facilitator participate in/facilitate events at other company locations and value streams or even at other companies.

During this development period, the facilitator should be exposed to a variety of kaizen scopes, complexity, tool application, and team and management dynamics. Some events will be more successful than others; but each one will impart a lesson(s) and provide an opportunity to become a better facilitator.

The facilitator eventually will learn how to teach more and direct less, and discern when it is best to accept (letting the teams learn from the process and their own mistakes) and when to demand. With this comes a much deeper understanding and appreciation for his role as a change agent, influencing and educating leadership at many levels.

NEXT STEPS

So, what is next for the enterprise that seeks to be kaizen-ready? Certainly, after gaining a significant level of proficiency and a commensurate level of lean transformation, the company can begin to indoctrinate its suppliers and customers (and even the customers' customers). This extends the scope and impact further upstream and downstream of the value stream. Clearly, from the perspectives of partnership and pragmatism, it is a win-win.

Most important, however, is for the enterprise to get progressively better at applying kaizen, both system driven and principle driven. True kaizen event effectiveness can only be achieved if the following characteristics are thoroughly, routinely, and profoundly satisfied. Kaizen events must be:

- performed in accordance with the standard work embodied in the multi-phase approach;
- driven by the strategy deployment and value-stream imperatives;
- considered only in the context of the company's lean business system;
- done as required, and supported and sustained by effective lean transformation leadership; and
- applied within a model that thoughtfully transitions from being sensei dependent, to kaizen promotion officer dependent, to employee driven.

If these characteristics are embodied within an enterprise, then that enterprise has removed many of the technical and cultural barriers to achieving sustainable lean transformation success. Indeed, such a company has effectively *countered* the bold statement made in Chapter 1:

> *A company that does not possess and routinely exercise the capability to effectively target, plan, execute, and follow-through on its kaizen events, including the non-negotiable requirement to comply with the new standard work to sustain the gains, cannot and will not successfully transform into a lean enterprise.*

The next challenge? It is obviously to evolve from an established system-driven kaizen capability to a principle-driven kaizen culture—one that effectively and widely leverages events, projects, and countless daily kaizen activities. Only then will kaizen truly be everyone's job, everyday. Kaizen!

SUMMARY

- The transition to kaizen readiness is enabled by:
    1. sensei guidance,
    2. effective transformation leadership, and
    3. solid short-term wins.

- Sensei guidance is a long-term investment. As the organization moves forward on its lean journey, it shifts from using the sensei predominately for kaizen event facilitation and cursory training to more substantial technical and transformation leadership coaching.

- One component of effective transformation leadership is the KPO organization. Organizations often do not grasp the urgency for KPO development because of outstanding kaizen events and sustained substantial improvements in the targeted value streams. The urgency is there, not because the results are not significant, but rather that they could be even better.

- The roadmap for KPO development includes:
    1. defining the KPO organization, including its structure, kaizen promotion officer roles and responsibilities, key result areas, and the requisite core and technical competencies.
    2. selecting and developing kaizen promotion officers—the best candidate identification and assessment tool is kaizen event performance. Development of kaizen promotion officers occurs through rigorous application, study, observation, mentorship, networking, benchmarking, volunteerism, and by "being a sponge."

- The KPO organization is an effective means for developing high-potential employees into excellent lean leaders who can later rotate into other key positions within the company.

- After the organization gains proficiency and experience in lean transformation, it can reach out to its suppliers and customers.
- Overall lean transformation requires the organization to target, plan, execute, and follow-through on its kaizen events, including the non-negotiable requirement to comply with the new standard work to sustain the gains. Without this ability, organizations will not successfully transform into lean enterprises.
- For true lean transformation, an enterprise must evolve from a system-driven kaizen capability to a principle-driven kaizen culture.

REFERENCES

"SME/AME/Shingo Prize Lean Certification Mentoring Guide." 2006. Dearborn, MI: Society of Manufacturing Engineers, version 2.0-04-2006, p. 1.

Womack, James P. and Jones, Daniel T. 1996. *Lean Thinking: Banish Waste and Create Wealth within Your Corporation*. New York: Simon Schuster, p. 270.

BIBLIOGRAPHY

Chiapppinelli, Chris. 2007. "Lean Benefits Misunderstood, Survey Finds." *ma News*, April 14. Accessed 6/19/2009 from Internet, http://www.managingautomation.com/maonline/news/read/Lean_Benefits_Misunderstood_Survey_Finds_30007.

Deming, W. Edwards. 1986. *Out of the Crisis*. Cambridge, MA: Massachusetts Institute of Technology.

Hirano, Hiroyuki. 1990. *5 Pillars of the Visual Workplace: The Sourcebook for 5S Implementation*. Portland, OR: Productivity Press.

Hirano, Hiroyuki. 1990. *JIT Implementation Manual: The Complete Guide to Just-In-Time Manufacturing*. Portland, OR: Productivity Press.

Huntzinger, Jim. 2005. *The Roots of Lean, Training within Industry: The Origin of Japanese Management and Kaizen*. Accessed 6-19-2009 from Internet, http://www.lean.org/admin/km/documents/39c9a597-5a72-4532-b046-ef0cfccbaa64-Roots%20of%20Lean%20-%20TWI.pdf.

Imai, Masaaki. 1997. *Gemba Kaizen: A Commonsense, Low-Cost Approach to Management*. NY: McGraw-Hill.

Japan Management Association, Lu, David J., ed. 1989. *Kanban: Just-in-Time at Toyota*. Portland, OR: Productivity Press.

Kotter, John P. 1996. *Leading Change*. Boston, MA: Harvard Business School Press.

Lean Enterprise Institute. 2003. *Lean Lexicon: A Graphical Glossary for Lean Thinkers*. Brookline, MA: The Lean Enterprise Institute, Inc.

Mann, David. 2005. *Creating a Lean Culture: Tools to Sustain Lean Conversions*. NY: Productivity Press.

Shingo Prize Model, 3rd edition. 2009. Accessed 6/19/09, http://shingoprize.org/files/uploads/TheShingoPrizeModel-ApplicationGuidelines0209.pdf. Logan, UT: Utah State University, Jon M. Huntsman School of Business.

Shook, John, and Rother, Mike. 1998. *Learning to See: Value Stream Mapping to Add Value and Eliminate Muda*. Brookline, MA: Lean Enterprise Institute.

Sobek, Durward K. II, and Smalley, Art. 2008. *Understanding A3 Thinking: A Critical Component of Toyota's PDCA Management System*. Boca Raton, FL: Productivity Press.

SME/AME/Shingo Prize Lean Certification Body of Knowledge, Version 3.0. 2008. December. Accessed 6/19/09, http://www.sme.org/downloads/cert/lean/facilitators_guide/TabT-LeanCert_BOK.pdf.

SME/AME/Shingo Prize Lean Certification Mentoring Guide. 2006. Dearborn, MI: Society of Manufacturing Engineers, version 2.0-04-2006. Accessed 1/4/09, http://www.sme.org/downloads/cert/lean/mentor_guide.pdf.

Stolovitch, Harold D., and Keeps, Erica J. 2006. *Telling Ain't Training*. Alexandria, VA: American Society for Training & Development.

Womack, James P., and Jones, Daniel T. 1996. *Lean Thinking: Banish Waste and Create Wealth in Your Corporation*. NY: Simon & Schuster.

GLOSSARY

This glossary is neither intended to be comprehensive nor to provide exhaustive definitions of key terms within the Fieldbook. Refer to other texts for more substantial glossaries of lean terms (see Lean Enterprise Institute 2003).

A

A3 report. A one-page (originally the international paper size A3, from which it derives its name) *plan-do-check-act* storyboard designed to facilitate problem-solving and sharing of information. The format typically reflects a problem statement or performance gap, brief analysis of the current condition, target condition, and an implementation plan to achieve the target. A3 thinking is reflected in the pre-event area profile and *improvement idea* forms.

C

countermeasure. An action or planned action intended to address the root cause of a specific issue or problem. To facilitate the identification, assignment, and status tracking of countermeasures, a standard form or flip chart is used with the following fields: problem or issue, action to be taken, assigned person, due date (and time), and completion status.

D

daily kaizen. These are small, process- or point-focused, continuous improvement activities conducted by engaged and enabled employees in their everyday work. Principle-driven lean enterprises apply a combination of kaizen events, projects, and daily kaizen. Daily kaizen opportunities (problems) are readily identified by workers using simple, robust lean management systems and by pragmatic comparison of the current state with the envisioned ideal state. Employees, as individuals and within teams, apply common sense and learning developed in kaizen events, training classes, and direct application as they engage in PDCA through the use/execution of actionable, low bureaucracy suggestion systems, mini-kaizen events, kaizen circle activities, "just-do-its," and the like.

define, measure, analyze, implement, and control (DMAIC). *Six sigma's* basic problem-solving methodology, in which a team: 1) defines a project's purpose and scope, 2) measures to determine the extent of the problem and identify its location(s) or source(s), 3) analyzes to ultimately identify root causes, 4) implements improvements or *countermeasures* to address root causes, and 5) controls by means of solution evaluation, standardization to maintain the gains, and formulation of steps for future improvements.

F

"fresh eyes." A term used to describe participants who are not intimately familiar with the *kaizen event's* area of focus. These individuals are typically unencumbered by any biases, have an open mind, and can easily ask the questions that many fail to ask, such as, "Why do we do that?" Each kaizen team should have one or more of these individuals.

G

gemba. A Japanese term that means "actual place," it is the place where the work is being done and at which kaizen activities should be directed.

genchi genbutsu. A Japanese term meaning "go and see for yourself." This reflects an orientation or attitude in which the *lean* practitioner habitually goes to the *gemba* to directly observe and, therefore, understand the situation.

H

heijunka. A Japanese term, roughly translated as "levelization." It is a foundational element of the *Toyota Production System*, facilitating the level and paced production or processing of both mix and volume over a fixed time.

I

improvement idea. A substantive idea implemented to help drive the team toward achievement of the kaizen targets. Typically there are a handful of high-impact ideas during a *kaizen event*. The team should capture the idea and the simple, underlying scientific thinking in an improvement idea form, which includes a description of the problem, description of the steps taken, the results, and a portrayal of the condition before and after the improvement.

initial strategy. The direction or plan of attack followed by the team, the initial strategy first and foremost focuses on the means and methods to conduct the initial observation of reality. It is dictated predominately by the kaizen scope, targets, realistic foreknowledge and understanding (based upon current situation data), and the duration of the *kaizen event*.

J

jidoka. This is the second of the two traditional *Toyota Production System* "pillars." It is a Japanese term meaning "automation with a human mind," and reflects the application of intelligent workers and machines that possess the ability to detect abnormal conditions and then apply appropriate *countermeasures*. One central aspect is line-stop jidoka, in which a process is stopped to contain the defect and then immediately address the root cause(s).

"just-do-its." These are improvements that should be realized through straightforward action, without the need for *kaizen events*, elaborate analyses, or projects. "Just-do-its" include policy changes, "no-brainer" capital expenditures, etc.

just-in-time (JIT). The first of the two traditional *Toyota Production System* "pillars," it is a customer-oriented, least-waste system for making or processing what is needed, when it is needed, in the amount needed, where needed. It is composed of three elements, supported and enabled by *heijunka*: 1) continuous flow, 2) pull system, and 3) *takt time*.

K

kaikaku. The Japanese term for radical improvement, it is in contrast to the small, continuous incremental improvements embodied by kaizen. Ironically, multi-day *kaizen events* often drive kaikaku.

kaizen. The Japanese term meaning "continuous improvement," it is used as both a noun and a verb. As a noun, the term "kaizen" is shorthand for a *kaizen event* or a smaller discrete kaizen activity, while "kaizens" (plural) is used for multiple kaizen events. As a verb, it represents the action of continuous improvement.

kaizen circle activity. Also known as a small group activity or quality circle, this is a primary strategy for daily kaizen. Kaizen circle activities are focused and facilitated team-based and team leader led activities that address a specific problem or opportunity (for example, develop a mistake-proofing device or procedure). Project-like in nature, they follow the PDCA rigor, are sponsored by a manager, and usually conducted over a period of weeks, requiring the team to meet routinely over that time (for example, one hour per week).

kaizen event. A focused, multi-level, cross-functional, team-driven continuous improvement activity in which *plan-do-check-act* (PDCA) and *standardize-do-check-act* (SDCA) rigor and *lean* thinking are applied to achieve targeted improvements. Kaizen events are also known as kaizen blasts, kaizen workshops, continuous improvement events, and lean events.

kaizen event area profile. A tool/template that facilitates the necessary kaizen event pre-planning rigor, including the identification and articulation of event scope, preliminary targets, linkage to strategic imperatives, event dates, team leader and team members, current situation and problems, etc.

kaizen event sequence. The core elements and nominal sequence within a kaizen event's execution phase. There are seven elements: 1) *kick-off meeting*, 2) pre-event training, 3) *kaizen storyline*, 4) *team leader meeting* process, 5) kaizen work strategy, 6) *report-out*, and 7) recognition and celebration. Elements 3, 4, and 5 are essentially conducted in parallel.

kaizen event multi-phase approach. The standard kaizen event management methodology, which is comprised of the following phases:

1. strategy—an at least annually refreshed activity that identifies and articulates the measurable high-leverage kaizen opportunities within the context of strategy deployment *x*-matrices and Gantt charts, value stream improvement plans, process improvement plans, and A3 reports.

2. pre-event planning—a phase that formalizes event selection and definition, effects communication to the appropriate stakeholders, prescribes and effects acceptable pre-work, and ensures that event logistics are addressed. This phase should be launched at least 20 business days before the event kick-off.

3. event execution—this phase represents the typical 3- to 5-day, team-based activity encompassing the seven-part kaizen event sequence.

4. event follow-through—the final phase, it focuses on completing the kaizen newspaper items and ensuring event improvement sustainability by means of things like leader standard work implementation and a post-event audit.

kaizen newspaper. A means for recording and tracking the status of critical *countermeasures* that remained unfinished at the end of a *kaizen event*. Typically, a standard form is used with fields for the problem or issue, specific countermeasure, assigned person, due date, and completion status. Nominally, such items should be closed out within 30 to 40 days after the event.

kaizen promotion office (KPO). The KPO represents the *lean* function for an enterprise. It is typically the dedicated, decentralized, multi-level (corporate, business unit, and value stream) lean resource group whose job responsibilities include kaizen event management, formulation of the company's lean business system, and curriculum development and training. The KPO is also known by other names, including lean office, lean promotion office, *just-in-time* (JIT) promotion office, continuous improvement office, and operational excellence.

kaizen storyline. This is the generic, *plan-do-check-act* (PDCA) -based execution roadmap for the kaizen team. It is output oriented, supported by a number of standard forms and templates, facilitates the visual storyboarding of the kaizen process and, ultimately yields the necessary elements and flow for the final *report-out*. There are eleven basic parts of the storyline: 1) scope and team, 2) kaizen targets, 3) *takt time* (customer requirements), 4) *initial strategy*, 5) *pre-kaizen situation*, 6) *countermeasures* and revised strategy, 7) implemented *improvement ideas*, 8) validated *post-kaizen situation*, 9) performance versus targets, 10) *kaizen newspaper*, and 11) *lessons learned*.

kaizen work strategy. This is the kaizen team's plan of action throughout the event as supplemented by tactics to facilitate team effectiveness. The strategy is initiated in the pre-event planning process and refreshed throughout the event. The team leader has primary ownership for its formulation and execution and routinely receives input and guidance by way of the *team leader meetings* as well as informal input from kaizen promotion officer and *sensei*. The generic strategy roadmap is the *kaizen storyline*.

kick-off meeting. The first step within the kaizen event execution phase, it is much more than a ceremonial meeting of senior leadership, visitors, and kaizen team members. The kick-off is a forum for a brief greeting and orientation; expression of commitment, expectation, and impact; team leader explanation of team scope, target, and initial strategy; *sensei* remarks and direction; and segue into pre-event training.

L

leader standard work. A *standardize-do-check-act* (SDCA) tool and the first of the four *lean management system* elements, it is a simple but powerful way to lock in kaizen gains. Leader *standard work* specifies day-to-day tasks and audit steps for the various leadership levels within the *value stream* and key functions. The audits are designed principally to assess, through the assistance of visual controls, process adherence and performance. The rigor of the leader standard work requires identification and recording of abnormal conditions (lack of process adherence and/or performance), related root causes, and *countermeasures*. Completed leader standard work is used as feedstock for the daily accountability process and to identify personal development opportunities.

lean. Virtually all glossaries wisely shy away from attempting a definition. This is due to the simultaneously multi-dimensional and holistic nature of lean. In other words, no definition is sufficient. Nevertheless, here is an attempt . . . lean is holistic business system, comprised of principles, tools, and techniques, which, when properly applied, facilitates the creation and delivery of the most *value*, as defined by the customer, while consuming the fewest resources.

Lean Certification. For the purpose of the Fieldbook, this represents the jointly sponsored Society of Manufacturing Engineers (SME), Association of Manufacturing Excellence (AME) and *Shingo Prize* Lean Certification. The Lean Certification is a multi-level knowledge and demonstrated application and mentorship-based credential founded upon the Shingo Prize model and related Lean Certification body of knowledge. Refer to www.sme.org/certification for further information.

lean management system (LMS). A critical system for sustaining kaizen gains through process adherence and process performance, the multi-level lean management system is comprised of *leader standard work*, visual controls, a daily accountability process, and leadership discipline. The LMS is a component of the *lean performance system*.

lean performance system. The author's term for the synergistic combination of *strategy deployment*, *value stream improvement planning*, and the *lean management system*. Lean leaders use the lean performance system tools, rigor, and its *plan-do-check-act* (PDCA) and *standardize-do-check-act* (SDCA) methodologies to drive breakthrough and day-to-day improvement, direct kaizen event selection, and develop the proper culture within the organization.

lessons learned. Knowledge absorbed by the team, individually and collectively, during the *kaizen event*. It usually encompasses insight into the kaizen process itself, group dynamics, target processes, technology, and team impact. The lessons, after brief reflection, are often captured by the team in a simple, facilitated session just prior to the *report-out*.

M

muda. The Japanese term for "waste," it represents any non-value-adding activity which, by definition, consumes resources but does not generate *value* for the customer. Muda is often distinguished as type 1 or type 2, the former representing waste that cannot be reasonably eliminated in the near term (for example, incidental work or work required due to regulatory requirements) and the latter, work that can be relatively easily eliminated through kaizen. Also see *seven wastes*.

mura. The Japanese term for "unevenness," it is a process typified by gyrations in demand, which in turn causes pulsing and the starting, stopping, hurrying, and idling of resources.

muri. The Japanese term for "overburden" or "strain," it represents unreasonable demand, which drives people, processes, and equipment beyond their capacities relative to throughput and duration, often resulting in availability and quality issues.

P

parking lot. Tool to briefly capture issues and concerns outside the kaizen scope and/or reasonable authority of the team, thus enabling the team to quickly move on to more relevant and actionable matters. The parking lot itself is typically a flip chart that reflects a description of the issue/problem/opportunity, recommended *countermeasure*, and recommended sponsor or owner. Parking lot items are reviewed in the event follow-through phase.

plan-do-check-act (PDCA). The scientific improvement process articulated by Walter Shewhart and later taught in Japan by W. Edwards Deming.

- Plan—establish the goals for the targeted process and identify the required changes (improvements) to achieve the goals.
- Do—implement the changes.
- Check—assess whether and how the executed plan delivered the intended results.
- Act—depending upon the results from the prior step, standardize and stabilize the improvements to sustain the improvements or begin the cycle again.

post-kaizen situation. The eighth part of the *kaizen storyline*, it represents the newly improved target process, previously the pre-kaizen situation transformed by implementation of the team's *countermeasures*. The post-kaizen situation should be documented and supported by standard work and visual controls, and validated by the kaizen team's direct observation.

pre-kaizen situation. The fifth part of the *kaizen storyline*, it represents the current condition as directly observed by the kaizen team and documented by tools such as time observation forms, spaghetti charts, standard work combination sheets, percent load charts, and process maps.

product family analysis matrix. A medium used to facilitate the identification of product (or service) families for the purpose of *value stream analysis* and improvement planning, it reflects products or services and the processes by which those products or services are created/delivered as well as the related quantity or frequency. The product family analysis matrix is also referred to as a product family matrix, process routing matrix, or product quantity process (PQPr) matrix.

process improvement plan. A formal plan that specifies the improvements, by means of the *kaizen events*, projects, and *"just-do-its"* captured in the kaizen "bursts," which are required to transform the mapped current state process to the mapped future state process. The process improvement plan, a primary driver of kaizen event selection, should identify the measurable goals and summary-level action steps for each improvement as well as the timing and the owner.

process mapping. A technique to: 1) document the current state of a process to identify issues and opportunities, 2) develop a least (lesser) waste future state with identified "kaizen bursts," and 3) specify a *process improvement plan*. Process mapping is often applied to long lead-time administrative processes, using forensic data such as files and transactions to provide insight into the current reality as a proxy for direct observation. It highlights hand-offs between people/function, quality checks, queue time/inactivity, etc.

R

report-out. This is the element of the kaizen event execution phase just prior to recognition and celebration. During the report-out, the kaizen team succinctly presents its journey and accomplishments within the context of the *kaizen storyline*. The audience should be comprised of senior leadership, outside visitors, other kaizen event teams (if a multi-team event), facilitators, and *sensei*, and, if possible, kaizen event affected employees. At the conclusion of the report-out, the sensei will provide some brief feedback, followed immediately by the senior leader/sponsor's closing comments.

S

sensei. The Japanese term for "teacher," it is a term often applied to master *lean* practitioners who facilitate the kaizen event process and teach willing students lean tools, techniques and, most importantly, principles. The sensei's preferred means for knowledge transference is learning by doing. The term "coach" is sometimes used in place of "sensei."

seven wastes. These are the original categories of waste articulated by Shigeo Shingo.

While they were traditionally associated with the wastes endemic within mass production, the seven wastes are present, at one level or another, within virtually all industries and value streams (also see *muda*):

1. overproduction—production that exceeds or is made ahead of customer requirements;
2. waiting—delays or idle time;
3. transportation—unnecessary physical movement of materials or information;
4. processing—unnecessary, excessive, or incorrect processing;
5. stock—materials and supplies in excess of what is needed;
6. motion—excessive, unnecessary or non-ergonomic motion; and
7. defects—rework and scrap.

Shingo Prize. The Shingo Prize for Operational Excellence, named after Shigeo Shingo, "was established in 1988 to promote awareness of *lean* manufacturing concepts and to recognize companies that achieve world-class manufacturing status around the globe." (Shingo Prize 2009) The scoring model and related criteria are fourfold: 1) cultural enablers, 2) continuous process improvement, 3) consistent lean enterprise culture, and 4) business results. Candidates are evaluated based upon the content of submitted achievement reports and, if selected, site visits.

six sigma. A multi-faceted quality management system that, when properly applied, is synergistic with *lean* and *kaizen*. It is constitutive of the following elements or characteristics:

- philosophy, which espouses customer focus and process variation reduction through the identification and control of critical inputs;
- methodology—*define, measure, analyze, implement, and control* (DMAIC) represents the primary six sigma methodology;
- toolkit—six sigma applies an array of tools and techniques to facilitate the *DMAIC* methodology, including: cause-and-effect diagrams, failure mode and effects analysis, statistical process control, hypothesis testing, components of variation, and designed experiments.
- goal—six sigma, from the perspective of process capability measurement (meaning six standard deviations between the process mean and either specification limit), reflects a process that yields no more than 3.4 defects per million opportunities.

standard work. A fundamental *lean* tool, it explicitly defines and communicates the current best practice (least waste way) for a given process that is dependent upon human action. Also known as "standardized work," it provides a routine for consistency, relative to safety, quality, cost, and delivery, and serves as a basis for improvement. Standard work is comprised of three basic elements: 1) *takt time*, 2) work sequence, and 3) standard work-in-process.

standardize-do-check-act (SDCA). A scientific approach used to sustain improvements through standardization.

- Standardize—develop standards for a specific process.
- Do—apply the standards.
- Check—assess whether the standards are sufficient and/or if there is a lack of adherence to the standard.
- Act—depending upon the results from the prior step, make adjustments/improvements to the standardized process and/or put *countermeasures* in place to address the deficit in process adherence.

strategy deployment. A management process in which an enterprise captures its critical few strategic imperatives, typically in terms of 3- to 5-year measurable objectives, "translates" them for each application area, and deploys them vertically and horizontally throughout the organization to the appropriate points of impact. Strategy deployment, also known as "hoshin kanri," strategic deployment, and policy deployment, is a primary driver of kaizen event selection.

T

takt time. The measure of the rate of customer demand, "takt" is a German word meaning "meter." Takt time is calculated by taking the available time of a given process for a given period and dividing it by customer demand for that period.

team leader meeting. A daily *plan-do-check-act* (PDCA) -based management process within the *kaizen event sequence* (the exception being the final day of the kaizen) in which team leaders briefly review their team's prior 24-hour accomplishments, strategy, and tactics for the next 24 hours, issues and barriers, etc. The audience includes local senior leadership, visitors, the kaizen promotion officer, and *sensei*. The process reinforces the rigor of the scientific *kaizen* process, holds team leaders accountable, engages management, facilitates the removal of barriers, ensures alignment with kaizen targets, facilitates and develops better *lean* thinking, and drives results.

Toyota Production System. The production system largely developed by Taiichi Ohno and employed by Toyota Motor Corporation to holistically facilitate the elimination of waste and thereby improve quality, reduce cost, and compress lead times. It is often visually portrayed in a "house" schematic that reflects interdependent elements such as *just-in-time*, *jidoka*, *heijunka* and *standard work*.

transformation leadership. Within the scope of the Fieldbook, this is characterized by lean leaders who consistently and effectively apply the transformation leadership model. The model harnesses both technical and emotional dimensions and is founded upon humility and respect for the individual. The technical aspect requires the lean leader to: 1a) apply kaizen event *standard work*, 1b) deploy *daily kaizen* throughout the organization, 2) apply a *lean performance system*, 3) establish and sustain a *kaizen promotion office*, 4) manage change, and 5) develop a personal lean competency. While technical deployment first moves the organization into a new way of being, it must be accompanied with lean leader competency, credibility, and emotional intelligence. The resultant dynamic will engender trust, expand tolerance for risk, and ultimately drive sustainable personal and organizational engagement, growth, and results.

"trystorm." The dynamic, real-time cycle of try-observe-improve-repeat, through which a kaizen team seeks to identify and validate the best *improvement idea*.

Training Within Industry (TWI). An approach comprised of three complementary programs, which were deployed to increase U.S. industrial productivity during World War II. Along with PDCA, TWI's job methods program represents the roots of *kaizen*. Within this program, supervisors are called to: 1) break down the targeted job, 2) question every detail, 3) develop a new, improved method, and 4) apply the new method.

V

value. The worth of a product or service as reflected by what the customer is willing to pay. Presuming adequate competition and relatively balanced supply and demand, the customer typically deems the first-time transformative (such as fabrication, assembly, or healing of a patient) or unique services (such as after-market troubleshooting, expert analysis, etc.) as creating value. This is in contrast to those activities that are not transformative or do not yield a unique service (the first time); in other words, those activities characterized by one or more of the *seven wastes*.

value stream. The actions, within a given product or service family, whether value-added or non value-added, necessary to bring that product or service to fruition, usually within the scope of new product introduction, customer inquiry to order, or order to remittance. The actions include the flow of material/products and information, and almost always cross functional boundaries.

value stream analysis (VSA). Kaizen of a given product or service family's material and information flow within a certain scope (generically, but not necessarily, apportioned between new product introduction, inquiry to order, and order to remittance). Conducted by a team comprised predominately of leaders who have responsibility for the product or service family, the value stream analysis yields a current state value stream map, a leaner future state value stream map, and a *value stream improvement plan*. The future state target accomplishment date typically does not exceed 12 months after the date of the VSA. VSA is also called "value stream mapping."

value stream improvement plan (VSIP). This is the third and most important output of a *value stream analysis* (the other two outputs are current and future state value stream maps). It is essentially a roadmap for achieving the future state within the prescribed target date. It is also a primary driver of kaizen event selection. The VSIP captures the detailed steps corresponding to the future state map's "kaizen bursts" and reflects the timing and *lean* tools as well as the assigned resources necessary for the completion of each step. The value stream manager is responsible for the overall plan's execution.

REFERENCE

Shingo Prize. 2009. Accessed from Internet, 6/19/09, http://shingoprize.org/htm/about-us/the-shingo-prize ...accessed.

APPENDIX A: BLANK FORMS

This Appendix of blank forms (see Table A-1) is provided to promote the use of standard work in your kaizen events. Feel free to improve the forms and tailor them to your needs (while remaining true to lean principles and kaizen event standard work). Additional reference materials are available on the Fieldbook's website, www.kaizenfieldbook.com.

Table A-1. Appendix A contents

Form Name	Appendix Page #	Fieldbook Reference
Strategy deployment x-matrix	230	Figure 3-4, p. 47
Value stream improvement plan	231	Figure 3-5, p. 48
Process improvement plan	232	Figure 4-6, p. 77
Master kaizen schedule	233	Figure 4-9, p. 81
Pre-event planning checklist	234	Figure 5-3, p. 89
Kaizen event area profile	235	Figure 5-4, p. 91
Kaizen team supply list	236	Table 5-6, p. 118
Kaizen event target sheet	237–238	Figure 5-12, p. 112; Figure 6-22, p. 171
Lean training module inventory	239	Table 6-3, p. 131
Countermeasure tracking form	240	Figure 6-9, p. 147
Kaizen newspaper	241	Figure 6-12, p. 153
Improvement idea form	242	Figure 6-11, p. 149
Team behavioral audit	243	Figure 6-17, p. 161
Bowling and Gantt chart	244	Figure 4-4, p. 75
Problem-solving A3 form	245	Figure 4-8, p. 80
Post-event follow-through checklist	246	Figure 7-3, p. 183
Kaizen event evaluation form	247	Figure 7-7, p. 191

Strategy Deployment X-matrix

Company name: _____
Team/level: _____
Year: _____

● = high correlation
O = moderate correlation

Appendix A: Blank Forms Kaizen Event Fieldbook

Value Stream Improvement Plan

Date prepared/revised: _____
Value stream manager(s): _____
Product/service family: _____
Date: _____
Business-level objective: _____

Index	Value Stream		Measurable Goal	Summary Level Action Steps	Impact	Cost	Monthly Schedule												Owner	Status
	Loop Name	Objectives					1	2	3	4	5	6	7	8	9	10	11	12		
																				25 50 75 100
																				25 50 75 100
																				25 50 75 100
																				25 50 75 100
																				25 50 75 100
																				25 50 75 100

Legend:

	Hi	Med	Low
Impact			
Cost	Low (Cost ≤ $___)	Med ($___ < Cost ≤ $___)	Hi (Cost > $___)

■ Plan ▨ On schedule ▨ Behind schedule

Process Improvement Plan

Date prepared/revised: _____   Process name: _____
Process manager: _____   Process objective: _____

Measurable Goal	Summary Level Action Steps	Impact	Cost	Monthly Schedule												Owner	Status
				1	2	3	4	5	6	7	8	9	10	11	12		
																	25 50 75 100
																	25 50 75 100
																	25 50 75 100
																	25 50 75 100
																	25 50 75 100
																	25 50 75 100
																	25 50 75 100
																	25 50 75 100
																	25 50 75 100

Legend:

Impact	Hi	Med	Low
Cost	Low (Cost ≤ $___)	Med ($___ < Cost ≤ $___)	Hi (Cost > $___)

■ Plan
▨ On schedule
▨ Behind schedule

Appendix A: Blank Forms

Master Kaizen Schedule

Date of last revision: _____ Business unit: _____ Scheduler: _____

#	Date	Location	Event Number	Event Description	Sponsor	Linkage SD	Linkage VSIP	Linkage PIP	Linkage Other	Description/Reference	KPO Internal	KPO External	Consultant (if Applicable)	Pre-event Planning Status 1 Event Select/Definition	Pre-event Planning Status 2 Communication	Pre-event Planning Status 3 Pre-work	Pre-event Planning Status 4 Logistics

☐ Started ☐ Completed

SD = strategy deployment
VSIP = value stream improvement plan
PIP = process improvement plan

233

Pre-event Planning Checklist

Event dates: _____   Process owner: _____

Team 1: _____   Team 3: _____
Team 2: _____   Team 4: _____

	Pre-event Categories		Required Activities	By Whom	Timing			Status %
					Recommended		Actual Completion Date	
					T-minus*	Date		
1	Event selection/ definition	1.1	Identify next kaizen opportunity and date		−30			25 50 75 100
		1.2	Verify schedule with facilitator/consultant		−45			25 50 75 100
		1.3	Identify team leader, prepare profile and submit to executive sponsor		−15			25 50 75 100
		1.4	Receive approved profile		−10			25 50 75 100
		1.5	Finalize team		−9			25 50 75 100
2	Communication	2.1	Develop kaizen week at-a-glance schedule		−10			25 50 75 100
		2.2	Conduct event orientation meeting with team		−8			25 50 75 100
		2.3	Apprise local management and out-of-town guests of event schedule, profile, etc. Ensure required management/supervisors will be present for team leader meetings and report-out		−8			25 50 75 100
		2.4	Provide specific communication to kaizen affected workers (support and targeted processes)		−7			25 50 75 100
		2.5	Provide general communication to the site		−5			25 50 75 100
3	Pre-work	3.1	Prepare target sheet		−9			25 50 75 100
		3.2	Identify/address perceived barriers		−8			25 50 75 100
		3.3	Prepare/perform data collection and analysis as appropriate		−5			25 50 75 100
4	Logistics	4.1	Reserve main conference room and team break-out rooms		−10			25 50 75 100
		4.2	Arrange for food and beverages		−5			25 50 75 100
		4.3	Procure/prepare recognition awards		−5			25 50 75 100
		4.4	Ensure items reflected on kaizen team supply list are present		−5			25 50 75 100
		4.5	Ensure petty cash availability, company credit card, etc.		−5			25 50 75 100
		4.6	Arrange security access/clearances for consultant and other visitors		−5			25 50 75 100

* = minimum number of business days from the event kick-off date

Kaizen Event Area Profile

Team name/#:		Profile revision date:		
Event description:		Event dates:		
Preliminary SMART objectives:		Team:		
			Name	Function
		TL		
		CL		
Location:		Support:		
Value stream/process:				
Value stream/process manager:			Name	Function
Executive sponsor:				
Strategy deployment/improvement plan linkage:		Facilitator:		
		Consultant:		
In scope:		Current situation and problems:		
Out of scope:				
Customer requirements (takt time):				
Process information:				

Appendix A: Blank Forms — Kaizen Event Fieldbook

235

Kaizen Team Supply List				
Pieces/ Units	Item		End of Event Inventory	Quantity Needing Replenishment
Stored within Plastic Storage Bin				
	Clipboards			
	Laminated copies of standard forms: 1) 5S audit sheet, 2) time observation form, 3) standard work sheet, 4) standard work combination sheet, 5) % load chart, 6) process capacity chart, 7) setup observation analysis sheet, 8) kaizen target sheet, 9) task list, 10) improvement idea form, 11) kaizen newspaper			
	Stopwatches			
	Pedometer			
	25-foot tape measure			
	Pencils (pre-sharpened)			
	White erasers			
	Pens			
	Flip chart markers (multi-colors)			
	Dry erase markers (multi-colors)			
	Dry erase eraser			
	Dry erase board cleaner			
	18-in. ruler			
	8-1/2 in. × 11-in. legal pads			
	Calculators			
	Stapler			
	Rolls of scotch tape in dispenser			
	Rolls of masking tape			
	Blank 8-1/2-in. × 11-in. overhead projector sheets			
	Paper clips			
	Rubber bands			
	Yellow sticky notes 3 in. × 3 in.			
	Orange sticky notes 3 in. × 3 in.			
	Green sticky notes 3 in. × 3 in.			
	Scissors			
	8-1/2-in. × 11-in. multi-color paper			
	11-in. × 17-in. multi-color paper			
	8-1/2-in. × 11-in. laminating pouches			
	11-in. × 17-in. laminating pouches			
	Sharpies (multi-colored)			
	Push pins			
	Adjustable 3-hole punch			
Not Stored within Plastic Storage Bin				
	Flip chart pads			
	Flip chart markers			
Shared Among Teams				
	Digital camera			
	Video camera			
	Label maker			
	Laminator			
	Measuring wheel			
	Kraft paper or white plotter paper			
	LCD projector (located in presentation room)			
	Overhead projector (located in presentation room)			
	Color printer (11-in. × 17-in. capable)			

Kaizen Event Target Sheet

Location: _____
Event: _____
Team lead: _____

Team name/#: _____
Date: _____

#	Key Measurement	Start	Target	Day 1	Day 2	Day 3	Day 4	End	IMPACT	
									Difference	% Improvement

Remarks:

Kaizen Event Target Sheet

Location: _____
Event: _____
Team lead: _____

Team name/#: _____
Date: _____

#	Key Measurement	Start	Target	Day 1	Day 2	Day 3	Day 4	End	IMPACT Difference	IMPACT % Improvement
1	5S score									
2	Productivity									
3	Throughput									
4	Crew size									
5	Lead time									
6	Cycle time (seconds)									
7	Setup/changeover time									
8	Defects									
9	Scrap									
10	Space (square feet, square meters)									
11	Inventory									
12	Work in process									
13	Parts' travel distance (feet, meters)									
14	Walking distance (feet, meters)									
15	Other:									

Lean Training Module Inventory

Module Title	File Name	Description/Purpose	Number of Slides	Training Duration (minutes)	Editor/Content Expert	Last Revision	Presenter Notes (Y/N)

Countermeasures Tracking Form

Team name/#: _____ Date: _____ Page: ____ of ____

#	Observation/Waste Identification Opportunity	Action to be Taken/Countermeasure	Person Responsible	Due Date	Percent Complete			
					25	50	75	100
					25	50	75	100
					25	50	75	100
					25	50	75	100
					25	50	75	100
					25	50	75	100

Appendix A: Blank Forms

Kaizen Newspaper

Team name/#:
Date:
Page ___ of ___

#	Problem	Countermeasure	Person Responsible	Due Date	Percent Complete			
					25	50	75	100
					25	50	75	100
					25	50	75	100
					25	50	75	100
					25	50	75	100
					25	50	75	100
					25	50	75	100
					25	50	75	100

Kaizen Event Fieldbook — Appendix A: Blank Forms

Improvement Idea Form

Kaizen scope: _____ Team name/#: _____ Completed by: _____ Date: _____

Description of Problem	Description of Steps Taken	Results

Before Kaizen

After Kaizen

Team Behavioral Audit

1. Rigor

| 1 | 2 | 3 | 4 | 5 | 6 | 7 | 8 | 9 | 10 |

No discipline, not on track — Following kaizen standard work and using effective work strategies

2. Urgency

| 1 | 2 | 3 | 4 | 5 | 6 | 7 | 8 | 9 | 10 |

Limited or inconsistent effort to get the right things done . . . no fire — Everyone is doing whatever it takes to satisfy kaizen targets

3. Participation

| 1 | 2 | 3 | 4 | 5 | 6 | 7 | 8 | 9 | 10 |

Few members dominate, some do not participate — Everyone contributes and is involved

4. Listening

| 1 | 2 | 3 | 4 | 5 | 6 | 7 | 8 | 9 | 10 |

>1 Person talks at a time, repetitions, interruptions, and side conversations — One person talks at a time; clarifying and building of ideas

5. Leadership

| 1 | 2 | 3 | 4 | 5 | 6 | 7 | 8 | 9 | 10 |

No attempts to bring team back on track and encourage equal participation — Team members intervene to keep team on track and manage equal participation

6. Decision quality

| 1 | 2 | 3 | 4 | 5 | 6 | 7 | 8 | 9 | 10 |

Team decisions based on opinion and inferior to individual assessments — Team expertise and decisions are data-driven and superior to individual judgments

7. Candor

| 1 | 2 | 3 | 4 | 5 | 6 | 7 | 8 | 9 | 10 |

Team members do not freely share thoughts — Team members speak openly

8. Fun

| 1 | 2 | 3 | 4 | 5 | 6 | 7 | 8 | 9 | 10 |

Rather have a root canal — ☺

Bowling Chart

Deliverable:

Link to strategy deployment:

Team:

Factors to consider:

Owner:

Date (last update):

Next Review:

Key measurement:

#	Action Steps	Owner	Assist	Annual Impact ($)	Jumping-off Point	20___ YTD											
							J	F	M	A	M	J	J	A	S	O	N	D
						Plan												
						Actual												
					Planned Date		J	F	M	A	M	J	J	A	S	O	N	D

Problem-solving A3 Form

Owner: _____
Date: _____

1. Theme:

2. Background

3. Current condition

4. Goal

5. Root cause analysis

6. Countermeasures

7. Effects confirmation

8. Follow-up action

Post-event Follow-through Checklist

Team name/#: _____
Team leader: _____
Report-out date: _____
Process owner: _____

Post-event Categories		Required Activities	By Whom	Final Day of Kaizen	Week 1, Day:					Post-kaizen Period			30-Day Audit
					1	2	3	4	5	Week 2	Week 3	Week 4	
1. Kaizen newspaper	1.1	Finalize newspaper.											
	1.2	Post newspaper.											
	1.3	Status newspaper items.											
2. Leader standard work (LSW)	2.1	Complete final LSW draft and validate linkages to visual controls.											
	2.2	Validate and approve LSW, allocate ownership to appropriate leaders, and integrate into existing LSW.											
	2.3	Deploy LSW ("go live").											
	2.4	Audit LSW. Identify/deploy necessary countermeasures.											
3. Audit	3.1	Determine current performance vs. reported kaizen achievements (including financial impact). Reconcile differences if any, assign countermeasures as required.											
	3.2	Conduct review of checklist status (all of above items), issues, etc. Identify/deploy necessary countermeasures.											
4. Kaizen event evaluation	4.1	Complete and submit evaluations.											
	4.2	Review and compile submitted evaluations (also reference lessons learned). Conduct follow-up discussions with submitters if clarification or additional information is required.										When appropriate	
	4.3	Reference formal and informal sensei feedback.											
	4.4	Conduct event post-mortem (start, stop, continue doing).											
	4.5	Identify necessary countermeasures. Review with appropriate people and assign.											
5. Kaizen summary	5.1	Complete kaizen event summary report.											
	5.2	Post/distribute summary report.											
6. Newsletter	6.1	Incorporate kaizen activity and results within regular lean/kaizen newsletter.										When appropriate	
7. Archiving	7.1	Submit summary report-out package to KPO.*											
	7.2	Quantify financial impact of improvements. Submit to KPO.											
	7.3	Catalogue and electronically file report-out package and financial impact estimate.											

Row groupings (left side): Sustainability (1, 2, 3); Event management improvement (4); Communication (5, 6); Record retention (7).

KPO = kaizen promotion officer, LSW = leader standard work
* Summary report-out package is to include at least the following: kaizen event area profile, kaizen target sheet (with results), kaizen newspaper, and kaizen lessons learned.

Appendix A: Blank Forms   **Kaizen Event Fieldbook**

Kaizen Event Evaluation Form

Participant name (optional): _____   Team: _____   Event date(s): _____

1. Including this event, I have participated in:
   - ○ 1 Event
   - ○ 2 Events
   - ○ 3 Events
   - ○ >3 Events

2. I participated in this event because I:
   - ○ Volunteered
   - ○ Was assigned

3. I was informed that I would participate in the event:
   - ○ >2 Weeks before
   - ○ 1-2 Weeks before
   - ○ <1 Week before

4. Were the event targets: Y   N
   - Clear? ○ ○
   - Important? ○ ○

5. Were you prepared?   Y ○   N ○

6. Did the team have everything it needed?   Y   N
   - Data ○ ○
   - Training ○ ○
   - Supplies ○ ○
   - Management support ○ ○
   - Other support (maintenance, IT, etc.) ○ ○

7. If there was anything missing, please describe:

   [_____]

8. Was your team successful?   Y ○   N ○

   If "No," why?

   [_____]

9. Would you like to participate in another event?:   Y ○   N ○

   If "No," why?

   [_____]

10. What would have made the event more effective?

    [_____]

Use other side if needed.

247

APPENDIX B:
DAILY KAIZEN—BEYOND THE EVENT

"The future comes one day at a time."
—Dean Acheson

The Fieldbook is principally about kaizen events and the related standard work as embodied within the multi-phase approach. However, the event, typically more kaikaku than kaizen in nature, does not, cannot, and should not replace truly small, cumulative continuous improvements. As such, it is useful to at least briefly explore the subject of daily kaizen.

The delivery system for small continuous improvements is daily kaizen. It follows then that when considering the practice of kaizen events versus daily kaizen, it is not an "either/or" decision; rather the two must pragmatically co-exist. Indeed, they are synergistic, with daily kaizen being ultimately applied most predominately.

As discussed in Chapter 2 (and in Figure 2-11), the Shingo Prize model and the Lean Certification Body of Knowledge appropriately recognize three levels of lean transformation and with it three levels of kaizen application:

1. *Tool driven*—management-planned kaizen for selected portions of a process without explicit linkage to strategic direction.
2. *System driven*—management- and engineering-planned kaizen linked to company strategies and value stream imperatives, with kaizen systematically facilitating the identification and elimination of waste, unevenness, and overburden.
3. *Principle driven*—"spontaneous continuous improvement via project, event, or 'just-do-it' approach; sponsored by management, engineering, work team, or worker. Kaizen activity is part of everyday work." (Shingo Prize model 2009)

Tool-driven kaizen is underdeveloped and, logically, ultimately ineffective. This is what was so unfavorably characterized in Chapter 4 as "kaizen without a cause"—drive-by kaizens, kamikaze kaizens, etc. The principle-driven level clearly represents the most mature stage of kaizen application and is reflective of both a system-driven use of kaizen events and robust and pervasive application of daily kaizen. As reflected in Figures B-1 and B-2, both kaizen

---

Daily Kaizen, a Definition

Daily kaizen is small, process- or point-focused, continuous improvement activities conducted by engaged and enabled employees in their everyday work. Principle-driven lean enterprises apply a combination of kaizen events, projects, and daily kaizen. Daily kaizen opportunities (problems) are readily identified by workers using simple, robust lean management systems and by pragmatic comparison of the current state with the envisioned ideal state. Employees, as individuals and within teams, apply common sense and learning developed in kaizen events, training classes, and direct application as they engage in PDCA through the use/execution of actionable, low bureaucracy suggestion systems, mini-kaizen events, kaizen circle activities, "just-do-its," and the like.

events and daily kaizen have a role to play in a lean transformation.

Multi-day kaizen events often represent the jump-start—the useful shock and awe for an enterprise's lean launch. Driving meaningful, often step-function, and hopefully sustainable performance improvements, they provide a dynamic training ground for kaizen participants, lean leaders, and many others. Events accelerate both learning curve and results and provide much needed momentum to the transformation. They establish a foundation of standard work and visual controls on which lean management systems can be built. This, in turn, creates a foundation for daily kaizen. Kaizen events thus serve as a necessary precursor and complementary partner of daily kaizen. Events, however, have their limitations.

Even the most aggressive kaizen event practitioners can cover only so much ground, considering that kaizen event participation opportunities for each employee are most likely in the single digits over a 24 to 36-plus month span. In contrast, with daily kaizen, the opportunities are, well . . . daily. A master kaizen calendar does not need to be checked to prove that there are many more business days than scheduled kaizen events. When the number of business days is multiplied by the number of workers, then the improvement opportunity count is staggering. This opportunity count extends not only to process improvement, but, perhaps more importantly, to the development and engagement of the workforce. Daily kaizen represents a grassroots process and employee capability improvement movement.

The expansive scope of practitioners leads some to refer to daily kaizen as "big K." This is in contrast to kaizen events or "little k." The distinction is meaningful and puts it in a logical context. Big K is not subservient to, nor is it replaced by, little k. This just would not make sense. However, many people fall into this trap

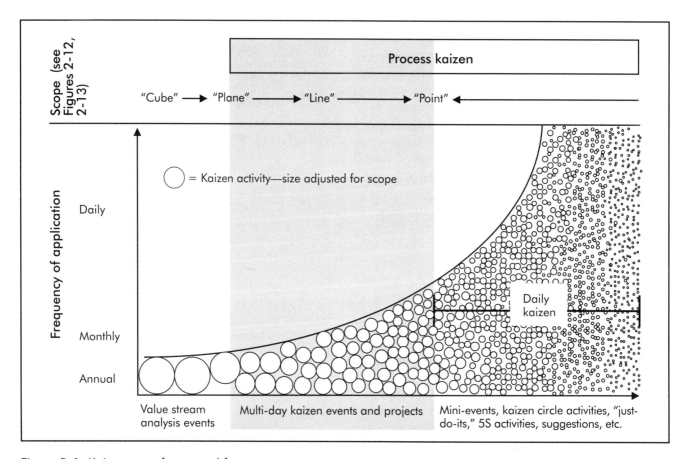

Figure B-1. Kaizen type, focus, and frequency.

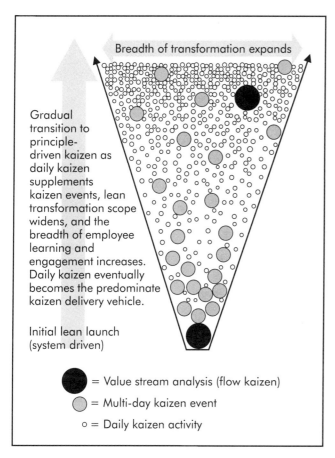

Figure B-2. A conceptual perspective of kaizen "mix" and sequence.

because lean launches typically start with kaizen events—first flow kaizen (value stream analysis) and then a number of process kaizen events as directed by the value stream improvement plan. What happens is that the newly indoctrinated organization sees the awesome impact of the events and leaders reasonably believe that this is a great way to transform the business. Daily kaizen then never establishes a foothold and kaizen is something that ends up being done only when there is a kaizen event scheduled. This inaction is the proverbial eighth waste of a person and a missed opportunity to truly transform an organization's culture into one of continuous improvement.

Overemphasizing kaizen events at the expense of daily kaizen can be analogously exemplified by the game of baseball. There is a phenomena in which people become overly enamored with home runs and disregard the need for what baseball folks call "small ball"—a run-producing strategy that relies on the execution of small incremental and cumulative activities. Small ball seeks to get and advance base runners through walks, singles, doubles, stolen bases, and sacrifice bunts and flies. The problem is that small ball, while not always flashy, is often the most reliable and certainly more team-oriented approach. It is imprudent to ignore small ball (daily kaizen) and instead "wait" for the home-run hitter to deliver a decisive blast (kaizen event). Heck, even Babe Ruth *only* hit a home run, on average, every 12.76 at bats! Further, home runs are clearly more impactful if there are players on base at the time. Analogously, many employees practicing daily kaizen also have this multiplier effect.

So, what exactly is daily kaizen? "Kaizen," as reflected in Chapter 2, is small continuous improvements. The meaning of "daily," in its strictest sense, is obvious. However, the combination of the two words yields a concept that transcends simply improving things everyday. Consistent with Imai's diagram (Figure 2-3), kaizen is everyone's job, from the very top of the organization (overhead) to the bottom (the value-adding worker). And it must be done with a frequency that eventually permeates everything at the process level that people, teams, and leaders of the organization do—think point kaizen (Figure 2-12). Figure B-3 provides further insight into daily kaizen.

Consistent with the kaizen system, daily kaizen is largely founded on standardize-do-check-act (SDCA) and plan-do-check-act (PDCA) thinking. In fact, the lean management systems established, maintained, and expanded throughout the lean launch and part of the kaizen event follow-through facilitate a perpetual comparison of actual versus standard. This comparison is made principally in the areas of process adherence (compliance with standard work) and process performance as part of a lean management system's combination of leader standard work, visual controls, and a multi-tiered daily accountability process. In such a situation, abnormal conditions or problems (process adherence and process performance gaps) can be identified by lean leaders and workers in a near real-time manner. These problems, perhaps more appropriately, "opportunities," can include things

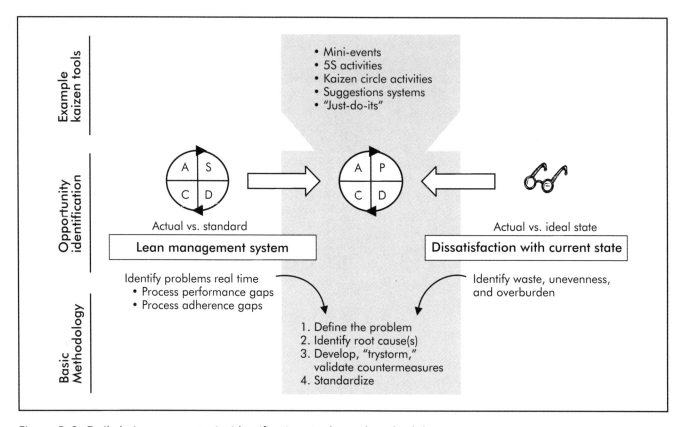

Figure B-3: Daily kaizen: opportunity identification, tools, and methodology.

like line stoppage due to machine breakdowns, defective parts, incomplete or inaccurate information, routine misses on hourly planned versus actual performance around start-up, customer complaints, supply levels that exceed those reflected in the (kanban-driven) supermarket, etc.

Not dissimilar to the dynamic of lean management system flagged problems is the routine identification of waste, unevenness, and overburden by trained, engaged workers who are constantly comparing the current state with a vision of an ideal state. Their observations are informed by an understanding of the seven wastes and supplemented with skills most likely developed by participating in kaizen events and kaizen circle activities. Examples of worker-identified problems include simple hassles or annoyances such as having to reach far and wide to retrieve an often used tool or reference material, difficulty in determining whether a material needs to be replenished or not, the frequent need to search out a supervisor to get a "rubber-stamped" approval, confusion as to which form is required for a particular process, etc. Worker-identified problems and opportunities can be focused by means of management-directed themes or campaigns. For example, management may sponsor a month-long focus on ergonomic improvements, accompanied by training, with the express intent to reduce the risk of repetitive motion injuries. Such a campaign may include an incentive for the team that contributes the most (implemented) improvements.

The effective identification of a problem is only the first step. It must be followed by the application of basic problem-solving methodology, in which:

1. The problem is clearly defined.
2. The root cause(s) is identified.
3. Countermeasures are developed, "trystormed," and validated.
4. The countermeasure becomes part of the new standard work.

The first two steps are facilitated by using tools and forms, with varying degrees of rigor, such as pre-event area profiles, A3 reports, improvement idea forms, simple employee suggestion forms/templates, 5-why analyses, mistake-proofing sheets, cause-and-effect diagrams, kaizen circle activity report forms, etc. Of course, formal basic problem-solving training (supplemental to lean-101-type training) should be provided to everyone within the organization and made real with encouraged and actual application at the gemba. True application will only occur if lean leaders provide the proper focus, resources (such as 5S supplies and kaizen promotion office and technician assistance), and time (for example, an hour a week for kaizen circle team activities).

The context or vehicle for the improvement cycle includes, but is not limited to:

- *Mini-events.* These events are measured in hours, not days. They are of limited and specific scope, while still applying the same basic PDCA approach of a multi-day kaizen event. An example would be the need to reduce the cycle time of a particular process or sub-process by 15 seconds so that it can repeatedly meet takt time.
- *Kaizen circle activities.* Also known as small group activities or quality circles (which many U.S. companies unsuccessfully tried to imitate from their Japanese counterparts during the 1980s), they represent focused and facilitated, team-based and team-leader led activities that address specific, meaningful problems or opportunities. Project-like in nature, a kaizen circle is sponsored by a manager, conducted over as many as 8 weeks, and requires the team to meet routinely over that time, for example, one hour per week. The activity typically culminates in a brief presentation to management, not dissimilar to a kaizen event report-out. An example activity is a mistake-proofing effort in which the team, applying the basic rigor of mistake-proofing:

  1. reviews the process steps within which the defect was generated,
  2. identifies the related red-flag conditions (for example, adjustments, lack of standards, multiple steps, etc.) that may cause the error/mistake, which then could cause the defect,
  3. identifies and validates the root cause(s),
  4. brainstorms mistake-proofing devices and/or procedures, and
  5. "trystorms"—validates and implements a device and/or procedure.

- *5S activity.* This foundation of lean is a grassroots vehicle for engagement and improvement. It offers countless, quick-hit opportunities to attack the seven wastes. For example, it takes only a few minutes and some 5S supplies to establish point-of-use storage for many small instruments and tools.
- *Suggestion system.* A well-designed lean management system will make use of a low bureaucracy, quick action (approval/rejection, assignment, and completion) suggestion system, and integrate its review and reference

---

Daily Kaizen is Done by Problem-solvers

A large portion of daily kaizen can be summarized as "find a problem, fix a problem, and prevent it from coming back" (by applying countermeasures that address the root cause[s]). Organizations have no shortage of problems, so this cycle is repeated in perpetuity. Like a muscle, repetitive problem-solving makes the worker and thus the enterprise stronger from the perspective of capability, engagement, and effectiveness. Similarly, four basic organizational capabilities have been identified that, if evolved to a proper level, deliver Toyota-like operational excellence (Spears 2005):

"1. Work is designed as a series of ongoing experiments that immediately reveal problems,
2. problems are addressed immediately through rapid experimentation,
3. solutions are disseminated adaptively through collaborative experimentation, and
4. people at all levels of the organization are taught to become experimentalists."

This is consistent with the scientific methods first introduced in Chapter 2, standardize-do-check-act (SDCA) and plan-do-check-act (PDCA).

within the tiered daily accountability process. It is within this SDCA context and the continual comparison of the current state with the ideal state, as supported by effective lean leaders and other resources, that workers will suggest a multitude of improvements. Depending upon the degree of difficulty and time requirements, often a suggestion can be implemented by the one suggesting it. In situations where there is need for procurement of materials/equipment, certain technical skills, or simply if implementation will take more than a few minutes, then it is often facilitated by supervisors, technicians, maintenance people, etc.

It is worthwhile to note that *any* type of kaizen, daily or event, in addition to exercising a proportionate level of PDCA/SDCA rigor, must be in keeping with change management needs. In other words, daily kaizen does not give license for people to run roughshod throughout the value stream applying "just-do-it" improvements. Certainly, there is more risk of this in the daily kaizen activities that have less formal facilitation than that used in mini-kaizen events or kaizen circle activities. There must be scientific thinking and there must be, in the spirit of humility and respect for the worker, an appreciation for the needs of the stakeholder—to understand the why, what, and how—and to be afforded a pragmatic opportunity to participate at the proper level and intensity. Lean leaders must see to that.

Kaizen events are an extremely critical part of any successful lean transformation. Properly done, they drive tremendous, sustainable results, priming the organization for further lean learning and deployment, and establishing a foundation and capability for daily kaizen. Kaizen events, and daily kaizen, with its grassroots process focus, frequency, and application, are the "dynamic duo," which cannot and should not be separated. To do so would be to the detriment of any lean transformation effort.

REFERENCES

Shingo Prize model. 2009. Accessed from Internet 4/20/09, http://www.shingoprize.org/Download/AwardInfo/BusinessPrize/ShingoPrizeModel.pdf, p. 4, 21.

Spear, Steven J. 2005. "Fixing Health Care from the Inside, Today." *Harvard Business Review*, September.

INDEX

10 + 1 principles, 22, 36
5S, 137, 170
5 why, 23, 101, 143

A

A3 report, 78, 80 (Fig. 4-8), 150, 221, 245
accountability, 51
adult learning, 128–129 (Fig. 6-3)
annual breakthrough objectives, 44
Ayers, Matthew, 53

B

Beran, Edward P., 151, 168
Berns, Evan M., 100
blank forms, 229–244
bowling chart, 244
breakthrough performance, 25

C

cause-and-effect diagram, 143 (Table 6-6)
celebration, 175–177
change,
   agent, 103
   leadership, 208–209
   management, 41, 43 (Table 3-1), 54–61, 203
   process, 43 (Table 3-1)
check sheets, 143 (Table 6-6)
checkpoint process, 46
Cho, Fujio, 107
Cicerrella, Carl M., 190
co-leader, 94–97
communication, 86–87, 108–111 (Fig. 5-11), 192–194

competency, 63, 207–212
concentration diagrams, 143 (Table 6-6)
consultant, 106–107 (Fig. 5-10)
core competencies, 95 (Fig. 5-5, 5-6), 102, 159–160, 207–210 (Fig. 8-6, 8-7, 8-8)
countermeasure, 142, 144–147 (Fig. 6-8, 6-9), 221, 240
credibility, 63–64
cross-talk, 169
cultural change, 61–63
cultural enablers, 21, 27–28
current state value stream map, 137
Cutler, James J., 63

D

daily kaizen, 4, 9, 13, 23, 25, 32, 41, 205, 221, 249–254
Danaher, 8, 63
data collection, 140–141
define, measure, analyze, improve, and control (DMAIC), 26, 132 (Fig. 6-4), 221
deliverables, 44, 46
Deming, W. Edwards, 18
deployment flow diagram, 137

E

Emotion Roadmap™, 57–58 (Fig. 3-8)
emotional intelligence model, 57–61, 64
emotional scope, 54–61
employee suggestions, 78
empowerment, 25, 28
environmental and safety systems, 28
event, 222
   30-day list, 152

event *(continued)*,
  area profile, 90–94 (Fig. 5-4), 222, 235
  audit, 187–188 (Fig. 7-5)
  coding, 193
  communication, 86–87, 108–111 (Fig. 5-11), 192–194
  consultant, 106–107 (Fig. 5-10)
  effectiveness, 3–4
  evaluation, 190–191 (Fig. 7-7), 247
  execution, 121–177
  facilitator, 105–107 (Fig. 5-10)
  feedback, 192
  financial impact, 189 (Fig. 7-6)
  flow, 91 (Fig. 5-4), 93, 135
  follow-through, 179–194 (Fig. 7-1, 7-2, 7-3), 243
  follow-up list, 152
  ground rules, 160
  impact, 74, 76, 78, 79, 83
  improvement ideas, 148–151 (Fig. 6-11), 242
  initial strategy, 72, 126 (Table 6-2), 163–165
  kick-off meeting, 123–127 (Table 6-1), 176, 223
  length, 36
  lessons learned, 154–155, 190–192
  logistics, 88, 116
  malpractice, 5 (Table 1-1)
  management, 107–108, 188, 194, 205–206
  master schedule, 232
  mission, 115 (Table 5-3)
  newspaper, 152–154 (Fig. 6-12), 181–184, 223, 241
  post-event, 187–188 (Fig. 7-5), 225, 246
  pre-event situation, 225
  pre-event training, 127–131 (Fig. 6-3, Table 6-3)
  pre-planning, 85–87 (Fig. 5-1, 5-2), 233
  pre-work, 111–116 (Table 5-5)
  pull, 29–31, 72–74 (Fig. 4-2, 4-3), 76, 78, 80 (Fig. 4-8)
  recognition and celebration, 175–177
  report-out, 172–175 (Fig. 6-24), 177, 225
  scheduling, 78–83 (Fig. 4-9), 88, 90, 109 (Table 5-2), 232
  scope, 92–93, 148 (Fig. 6-10)
  selection, 72, 79, 82 (Fig. 4-10), 88–107
  sense of urgency, 160–162 (Fig. 6-18)
  sensei, 8, 127, 192, 198, 217, 225
  sequence, 11, 121–123 (Fig. 6-1), 222
  standard work, 4, 8, 41, 123
  storyline, 132–155 (Fig. 6-4, 6-5, Table 6-4), 223
  strategy, 72–78 (Fig. 4-4), 146, 148, 163–165
  summary report, 192–193
  supply list, 118 (Table 5-6), 236
  takt time, 123–124 (Fig. 6-2), 226
  target sheet, 111–114 (Fig. 5-12), 171 (Fig. 6-22), 237–238, 242
  team behavioral audit, 161 (Fig. 6-17), 243
  team effectiveness, 158–172 (Fig. 6-14)
  team leader, 54
  team selection, 94–108, 119
  team training, 105
  timing, 79
  without a cause, 69
  work strategy, 164 (Fig. 6-19), 177, 223
executive A.D.D., 166, 199

F

facilitator, 105–107 (Fig. 5-10)
Fieldbook, 8–12, 42 (Fig. 3-2)
fish rotting, 39
flow (material and information), 32, 170
flow kaizen, 32, 91 (Fig. 5-4), 93, 135, 182
Ford, Henry, 19
forms, blank, 229–244
Foster, Jerry C., 41
"fresh eyes," 99–101, 221
functional managers, 54

G

gap closure, 76
gemba, 3, 22, 222
genchi genbutsu, 23, 135, 222
green targets, 113
"group think," 102

H

hansei, 155
heijunka, 34, 222
Hirano, Hiroyuki, 31, 33 (Fig. 2-12), 135
histograms, 143 (Table 6-6)
hoshin kanri, 44
humility, 22, 24, 61

I

Imai, Masaaki, 19–20 (Fig. 2-3), 32–34 (Fig. 2-13)
improvement, 24 (Fig. 2-7), 31–32
  idea, 148–151 (Fig. 6-11), 222, 242
  planning, 76–77 (Fig. 4-5, 4-6)

initial kaizen event, 72, 126 (Table 6-2)
initial strategy, 163–165, 222
Ishikawa diagram, 143 (Table 6-6)

J

Jake Brake, 8
Jeffrey, Richard A., 117, 167
jidoka, 28, 222
"just do it," 8, 25, 31, 70, 160, 222
just-in-time, 25, 28, 222

K

kaikaku, 13, 222
kaizen, 16–21, 36, 217, 222–223
   30-day list, 152
   audit, 187–188 (Fig. 7-5)
   barriers, 114–115 (Table 5-4)
   circle activity, 222
   communication, 86–87, 108–111 (Fig. 5-11), 192–194
   consultant, 10, 106–107 (Fig. 5-10)
   cultural enablers, 21, 27–28
   daily, 4, 9, 13, 23, 25, 32, 41, 205, 221, 249–254
   decision logic, 71 (Fig. 4-1)
   definition, 4, 13–14
   deployment, 26 (Fig. 2-9)
   enablers, 197
   evaluation, 190–191 (Fig. 7-7), 247
   event area profile, 90–94 (Fig. 5-4), 223, 235
   event coding, 193
   event effectiveness, 3–4
   event mission, 115 (Table 5-3)
   event pull, 73–74 (Fig. 4-2, 4-3)
   event selection, 79, 82 (Fig. 4-10), 88–107
   event sequence, 11, 121–123 (Fig. 6-1), 223
   event summary report, 192–193
   event target sheet, 111–114 (Fig. 5-12), 171 (Fig. 6-22), 237–238, 242
   facilitator, 105–107 (Fig. 5-10)
   feedback, 192
   financial impact, 189 (Fig. 7-6)
   flow, 32, 91 (Fig. 5-4), 93, 135, 182
   follow-through, 179–194
   follow-up list, 152
   ground rules, 160
   impact, 74, 76, 78, 79, 83
   improvement ideas, 148–151 (Fig. 6-11), 222, 242

kaizen (continued),
   initial event, 72, 126 (Table 6-2), 163–165
   kick-off meeting, 123–127 (Table 6-1), 176, 223
   length, 36
   lessons learned, 154–155, 190–192
   levels, 31–32, 34 (Fig. 2-13)
   logistics, 88, 116
   malpractice, 5 (Table 1-1)
   management, 107–108, 188, 205–206
   master schedule, 233
   methodology, 21, 24–26 (Fig. 2-9)
   multi-phase approach, 36–37 (Fig. 2-15), 223
   newsletter, 193
   newspaper, 152–154 (Fig. 6-12), 181–184, 223, 241
   office management, 205–206
   opportunism, 30–31
   philosophy, 21–22
   post-event audit, 187–188 (Fig. 7-5), 225
   post-event follow-through, 180–183 (Fig. 7-1, 7-2, 7-3), 246
   pre-event planning, 85–87 (Fig. 5-1, 5-2), 233
   pre-event situation, 225
   pre-event training, 127–131 (Fig. 6-3, Table 6-3)
   pre-work, 111–116 (Table 5-5)
   principles, 21–24 (Fig. 2-6), 70
   pull, 29–31, 72–74 (Fig. 4-2, 4-3), 76, 80 (Fig. 4-8)
   readiness, 197–218
   recognition and celebration, 175–177
   record retention, 193–194
   report-out, 172–175 (Fig. 6-24), 177, 225
   resources, 79, 88, 90
   scheduling, 78–82 (Fig. 4-9), 83, 88, 90, 109 (Table 5-2), 233
   scope, 92–93, 148 (Fig. 6-10)
   sense of urgency, 160–162 (Fig. 6-18)
   sensei, 8, 127, 192, 198, 217, 225
   SMART objectives, 92
   standard work, 4, 8, 41, 123
   storyline, 132–155 (Fig. 6-4, 6-5, Table 6-4), 223
   strategy, 72–78 (Fig. 4-4), 146, 148, 163–165, 230
   supply list, 118 (Table 5-6), 236
   system, 21–22 (Fig. 2-5), 32
   takt time, 123–124 (Fig. 6-2), 226
   targets, 114

kaizen *(continued)*,
   team behavioral audit, 161 (Fig. 6-17), 243
   team effectiveness, 158–172 (Fig. 6-14)
   team leader, 54
   team selection, 94–108, 119
   team training, 105
   timing, 79
   tools, 21, 26–27 (Fig. 2-10)
   universality, 4
   without a cause, 69
   work strategy, 164 (Fig. 6-19), 177, 223
kaizen promotion office(r), 6–7 (Fig. 1-2), 41, 54, 200–216, 223
   accountability, 209
   barriers to, 198–200
   career path, 212
   change leadership, 208–209
   change management, 203
   competency, 207–212
   curriculum development, 203–204 (Fig. 8-5)
   event management, 205
   facilitator training, 215 (Fig. 8-11)
   flexibility, 209
   group leadership, 208
   improvement, 206
   interpersonal understanding, 209
   key result areas, 203 (Fig. 8-4)
   lean business system, 8, 32, 35 (Fig. 2-14), 203–204 (Fig. 8-5)
   lean deployment improvement, 206
   learning, 214
   office management, 205–206
   officer selection, 213–216
   people development, 205, 209
   return on investment, 206–207
   roadmap, 198–200 (Fig. 8-2)
   roles/responsibilities, 202–206
   self-management, 209–210
   strategic orientation, 207
   structure, 200–202 (Fig. 8-3)
   technical competency, 210-211 (Fig. 8-9)
"kaizenology," 13
kick-off meeting, 123–127 (Table 6-1), 176, 223
Klesczewski, Robert L., 31
Kotter, John P., 41, 43 (Table 3-1)

L

leader standard work, 50, 184–187 (Fig. 7-4), 224

leadership, 24, 39–65, 208–209
lean, 197, 224
   barriers, 3
   business system, 8, 32, 35 (Fig. 2-14), 203–204 (Fig. 8-5)
   Certification, 9, 28–29, 130, 204 (Fig. 8-5), 209–212 (Fig. 8-9, 8-10), 224
   competency, 43
   consultant, 10, 106–107 (Fig. 5-10)
   deployment improvement, 206
   fakers, 41
   leaders, 55 (Fig. 3-7)
   leaders' must do, cannot fail list, 40, 72, 86, 125, 180, 200
   management system, 49–52 (Fig. 3-6), 78, 224
   newsletter, 193
   performance system, 41, 43–52 (Fig. 3-3), 224
   practitioners, 4–8 (Fig. 1-1)
   prerequisite for, 3–4
   return on investment, 188–189 (Fig. 7-6)
   training module inventory, 239
   transformation, 3–4, 39–40 (Fig. 3-1), 52–54
lessons learned, 154–155, 190–192, 224
logistics, 116
long-term scheduling, 78–82 (Fig. 4-9), 83, 88, 90

M

MacGyver, 23, 25, 162–163
master kaizen schedule, 233
Mayer, John, 57
measurement (success), 172
muda, 13–16, 148 (Fig. 6-10), 224
   categories, 14–15 (Table 2-1)
   countermeasures, 142, 144–147 (Fig. 6-8, 6-9)
   elimination math, 136 (Fig. 6-6)
   identification tools, 137–139 (Table 6-5), 141 (Fig. 6-7)
   root cause, 141 (Fig. 6-7), 143 (Table 6-6)
   types, 14, 37, 225–226
mura, 15–16, 224
muri, 15–16, 224
Murli, Joe, 165
must do, cannot fail list, 40

N

negative emotions, 56
no layoff policy, 30
non-value-added work, 14

O

observation, 135
O'Connor, Michael, 113
Ohno, Taiichi, 8, 19, 20, 23, 28, 63, 135
operations analysis table, 139
operator balance chart, 138
operator loading diagram, 138
opportunism, 30–31, 76, 78

P

Pareto charts, 143 (Table 6-6)
"parking lot," 142, 165–166 (Fig. 6-20), 224–225
people development, 28
perceived barriers, 114–115 (Table 5-4)
percent load chart, 138
performance gaps, 30
performance vs. kaizen targets, 152
plan-do-check-act (PDCA), 16–19 (Fig. 2-1), 21 (Fig. 2-4), 79 (Fig. 4-7), 163, 169 (Fig. 6-21), 225
plus delta activity, 165
point-to-cube continuum concept, 32–33 (Fig. 2-12)
post-event audit, 187–188 (Fig. 7-5)
post-event follow-through, 180–183 (Fig. 7-1, 7-2, 7-3), 246
post-kaizen situation, 151–152, 225
pre-event planning, 85–87 (Fig. 5-1, 5-2), 234
pre-event training, 127–131 (Fig. 6-3, Table 6-3)
pre-kaizen situation, 135–142, 225
pre-work, 111–116 (Table 5-5)
pro forma results, 152
process,
    capacity sheet, 139
    failure modes and effects analysis, 143 (Table 6-6)
    improvement plan, 74, 77 (Fig. 4-6), 225, 232
    mapping, 49, 137, 225
product family analysis matrix, 225
productivity improvement, 168
project management, 50
pull, 29–31, 34, 170

R

recognition, 175–177
record retention, 193–194
report-out, 172–175 (Fig. 6-24), 177, 225
resource scheduling, 79, 88, 90
respect, 22, 24, 27–28, 61

return on investment, 188–189 (Fig. 7-6)
Ringelmann, Maximilian, 98–99
risk, 64
Rizzo, John A., 90, 101, 127
Robbins, Craig, 61, 70, 213
root cause, 141–143 (Fig. 6-7, Table 6-6)

S

safety and environmental systems, 28
Salovey, Peter, 57
scatter diagrams, 143 (Table 6-6)
scientific thinking, 24–25 (Fig. 2-8)
scope creep, 98, 164–165
sense of urgency, 160–162 (Fig. 6-18)
sensei, 8, 127, 225
    feedback, 192
    proficiency, 8
    Socratic Method, 8
serial kaizen, 180
setup observation analysis work sheet, 139
seven wastes, 225–226
Shewhart cycle, 17–19 (Fig. 2-1, 2-2)
Shingijutsu, 8, 167
Shingo Prize, 27–29, 226
Shingo, Shigeo, 14
single-minute exchange of dies (SMED), 62
six sigma, 26, 226
SMART objectives, 92
social loafing, 98–99
Society of Manufacturing Engineers (SME), 9
solution, 144
specific, measurable, actionable, relevant, time-bounded (SMART) objectives, 92, 115
standard,
    forms, 140
    work, 4, 20, 41, 51, 170, 226
    work combination sheet, 138, 173 (Fig. 6-23)
    work-in-process (WIP), 9, 20
standardize-do-check-act (SDCA), 17 (Fig. 2-1), 19–21 (Fig. 2-4), 30, 151, 226
status quo, 23
status tracking, 169 (Fig. 6-21)
Stiles Associates, LLC, 53
storyboard, 132, 170
strategic initiatives, 44
strategy, 69–83
strategy deployment, 44, 46–47 (Fig. 3-4), 72–78 (Fig. 4-4), 75 (Fig. 4-4), 226, 230

superficial kaizen, 180
supervisors, 54
supply list, 118 (Table 5-6), 235
sustaining improvements, 180–188, 194
Suzumura, Kikuo, 107
system kaizen, 32

T

takt time, 4, 9, 20, 93, 123–124 (Fig. 6-2), 226–227
target sheet, 111–114 (Fig. 5-12), 237–238
team,
   behavioral audit, 161 (Fig. 6-17), 243
   effectiveness, 158–172 (Fig. 6-14)
   leader, 54, 94–97 (Fig. 5-5, 5-6), 159–160 (Fig. 6-15)
   leader meetings, 155–158 (Fig. 6-13), 177, 227
   presentation, 173 (Fig. 6-23)
   selection, 94–108 (Table 5-1, Fig. 5-5 through 5-9), 159–160, 207–210 (Fig. 8-6, 8-7, 8-8)
   technical competencies, 95–97 (Fig. 5-5), 101–102, 159–160, 210–211 (Fig. 8-9)
   training, 105, 127–131 (Fig. 6-3, Table 6-3)
technical change, 61–63
Thedacare, 103
Thompson, Bruce E., 53, 86
time observation form, 137
Toyota Production System, 8, 25, 28, 31, 34, 63, 227
training, 105, 127–131 (Fig. 6-3, Table 6-3)
   curriculum, 130–131 (Table 6-3)
   materials, 128
   module inventory, 130–131 (Table 6-3), 239
   Within Industry (TWI), 16–18 (Fig. 2-1), (Table 2-2), 227
transformation leadership, 24, 39–65 (Table 3-1, Fig. 3-2, 3-9), 54–61, 123, 203, 227
trust, 64
"trystorm," 150, 227

U

urgency, 160–162 (Fig. 6-18)

V

value, 14, 227
value-added work, 14
value stream, 227
   analysis, 227

improvement plan, 46, 48–49 (Fig. 3-5), 74, 228
leadership, 54
vision, 60
visual,
   controls, 51
   language, 140
   management, 170–172

W

waste, 13–16, 148 (Fig. 6-10)
   categories, 14–15 (Table 2-1)
   countermeasures, 142, 144–147 (Fig. 6-8, 6-9)
   elimination math, 136 (Fig. 6-6)
   identification tools, 137–139 (Table 6-5), 141 (Fig. 6-7)
   root cause, 141 (Fig. 6-7), 143 (Table 6-6)
   seven, 225–226
   types, 14
"wastology," 135
Wolfe, Charles, 54–55
work sequence, 20
work strategy, 158–172 (Fig. 6-19), 177
worker redeployment, 31
worker suggestions, 31

X

x-matrix, 46–47 (Fig. 3-4)